Fluid Mechanics and Machinery

Dr.R. Sivasubramanian, B.E., M.E., Ph.D.,

Principal,

Latha Mathavan Engineering College,

Kidaripatti, Madurai.

Published by

Fluid Mechanics and Machinery

Second Edition: March 2019

ISBN 978-93-86638-23-6

Author

Dr.R. Sivasubramanian

Bonfring
309, 2nd Floor, 5th Street Extension, Gandhipuram,
Coimbatore-641 012.
Tamilnadu, India.
E-mail: info@bonfring.org
Website: www.bonfring.org
Phone: 0422 4213231

Dedicated

To the loving

Parents, my family and my colleagues and friends

Preface

I am glad to present the book entitled, "A Text Book of Fluid Mechanics and Machinery" to the engineering students of Mechanical and Civil. The course contents have been planned in such a way that the general requirements of all engineering students are fulfilled.

During my long experience of teaching this subject to undergraduate engineering students for the past 15 years. I have observed that the students face difficulty in understanding clearly the basic principles, fundamental concepts and theory without adequate solved problems along with the text. To meet this very basic requirement to the students, a large number of the questions taken from the examinations have been solved along with the text in M.K.S. and S.I. Units.

The book is written in a simple and easy-to-follow language, so that even an average students can grasp the subject by self-study. At the end of each chapter highlights, theoretical questions and many unsolved numerical problems with answer are given for the students to solve them.

I am thankful to my colleagues, friends and students who encouraged me to write this book. I express my appreciation and gratefulness to my publisher for his most cooperative, painstaking attitude and untiring efforts for bringing out the book in a short period.

Though every care has been taken in checking the manuscripts and proof reading, yet claiming perfection is very difficult. I shall be very grateful to the readers and users of this book for pointing any mistakes that might have crept in. Suggestions for improvement are most welcome and would be incorporated in the next edition with a view to make the book more useful.

R. Sivasubramanian

<table>
<tr><th>Unit</th><th>Contents</th><th>Page No</th></tr>
</table>

UNIT I

FLUID MECHANICS AND MACHINERY

1.1. Introduction

Definition

Fluid Mechanics is the branch of science which deals with behavior of fluids at rest as well as in motion.

Fluid Mechanics deals with static, kinematics and dynamics aspect of fluids.

Fluid Static

The study of fluid at rest is called as Fluid Statics.

Fluid Kinematics

The study of fluid in motion where the pressure force are not considered is called as fluid kinematics.

Fluid Kinetics (Dynamics)

The study of fluid in motion where the pressure forces are considered is called as fluid kinetics.

Dimensions

(i) 1 m= 100 cm

    cm is converted into meter

(ii) 1 m= 1000 mm

    mm is converted into meter

(iii) Unit = Distance

    Distance= x (or) l in metre

(iv) Unit – diameter

    Diameter=d (or) D in metre

(v) Unit – Velocity

$$\text{Velocity (v)}= \frac{Dis\tan ce}{time} \text{ in m/s}$$

1 cm =	$\dfrac{1}{100}\,m$
1 mm =	$\dfrac{1}{1000}\,m$

(vi) Unit – Acceleration

$$\text{Acceleration (a [or] g)} = \frac{Velocity(v)}{time(t)} = \text{m/s}^2$$

(vii) Newton second law

F	=	ma
Force	=	Mass x Acceleration
	=	kg x m/s²
F	=	1 kg. m/s² =1 N

(viii) Difference between mass and weight

Sl. No.	Mass	Weight
1.	Without considering the gravity	With considering the gravity
2.	Symbol (m); Unit in kg	Symbol (w); Unit in m x g = kg x m/s² Weight force = 1 N

1.2. Properties of Fluids

1) Density (or) Mass Density

$$\text{Mass Density (or) Density }(\rho) = \frac{mass}{volume} = \frac{kg}{m^3}$$

2) Weight density (or) Specific weight

$$\text{Weight density (w)} = \frac{Weight}{volume} = \frac{Weight\ force}{Volume} = \frac{F}{V}$$

$$= \frac{m\ x\ a}{V} = \frac{kg}{m^3} x \frac{m}{s^2}$$

$$w = \frac{N}{m^3}$$

Note

(i) ρ - Density of water $\left(\rho_{H_2O} = 1000 kg/m^3\right)$

(ii) Acceleration due to gravity

$$\text{a (or) g} = 9.81 \text{ m/s}^2$$

3) Weight density w $= \dfrac{Weight\ force}{volume}$

$$w_{H_2O} \quad = \quad \frac{F}{V} = \frac{m \times a}{V} = \frac{m}{V} xa$$

$$= \quad \rho \times a$$

$$w_{H_2O} \quad = \quad 1000 \text{ kg/m}^3 \times 9.81 \text{ m/s}^2$$

$$w_{H_2O} \quad = \quad 9810 \text{ N/m}^3$$

4) Specific volume $\left(\dfrac{1}{\rho}\right) = \dfrac{Volume}{mass} = \dfrac{m^3}{kg}$

5) Specific gravity (or) Relative Density

$$s \quad = \frac{Weight\ density\ of\ liquid}{Weight\ density\ of\ s\tan dard\ liquid} = \frac{\rho_{liquid}}{\rho_{s\tan dard\ liquid}}$$

Note 1

1. Specific gravity of water (s) $\quad = \quad 1$
2. Specific gravity of mercury (s) $\quad = \quad 13.6$

Note 2

$$Volume = \quad 1 \text{ lit} \quad = \quad \frac{1}{1000} m^3 OR \quad 1000 cm^3$$

6) Viscosity or Dynamic Viscosity

Unit for dynamic viscosity $\quad = \quad$ Shear stress x Shear Strain

$$\tau \times \frac{du}{dy}$$

$$\tau \quad = \quad \mu.\frac{du}{dy}$$

$\mu \quad - \quad$ Dynamic viscosity

$$\mu \quad = \quad \frac{\tau}{\dfrac{du}{dy}}$$

$$\mu \quad = \quad \frac{\dfrac{F}{A}}{\dfrac{u-0}{y-0}} = \frac{F}{A} x \frac{y}{u}$$

$$\mu \quad = \quad \frac{Nm}{m^2.m} xs$$

Dynamic Viscosity (μ) $\quad = \quad \dfrac{NS}{m^2}$

Note

$$1 \text{ poise} = \frac{1}{10}\frac{NS}{m^2}$$

7) Kinematic viscosity

$$\text{Kinematic Viscosity } (\gamma) = \frac{Dynamic\ Vis\cos ity(\mu)}{Density(\rho)}$$

$$\text{Unit} \quad \gamma = \frac{NS}{m^2 \cdot \dfrac{kg}{m^3}} = \frac{kg.m.s}{s^2 .m^2}\frac{m^3}{kg}$$

$$\gamma = \frac{m^2}{s}$$

NOTE: 1 stoke = 10^{-4} m²/s

8) Bulk Modulus (K)

$$\text{Bulk Modulus (K)} = \frac{Increase\ in\ pressure}{Volumetric\ Strain} = \frac{dP}{\dfrac{-dV}{V}}$$

$$K = \frac{-VdP}{dV}$$

$$\text{Unit} \quad K = m^3 \frac{N}{m^2 x m^3}$$

$$K = \frac{N}{m^2}$$

9) Compressibility

$$\text{Compressibility} = \frac{1}{Bulk\ \bmod ulus(K)}$$

Unit is m²/N

10) Surface Tension

Surface tension is defined as a tension force acting on a surface of a liquid in contact with a gas such that the contact surface behaves like a membrane under the tension.

Case 1

Consider a small spherical droplet of a liquid of radius 'r'

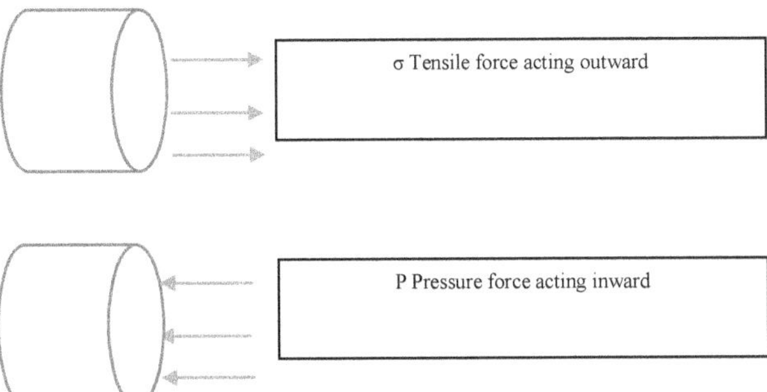

Figure 1.1: Liquid Droplet

The entire surface of the droplet, a tensile force due to surface tension will be acting.

Let,

$\sigma \rightarrow$ Surface tension of the liquid (N/m)

$P \rightarrow$ Pressure inside the droplet (N/m²)

$d \rightarrow$ Diameter of the droplet (m)

Step 1

Tensile force acting on the surface $\quad = \quad \sigma \times \pi\, d \qquad$ (N)

Step 2

Pressure force acting on the surface $\quad = \quad P \times \dfrac{\pi}{4} d^2 \qquad$ (N)

Step 3

For equilibrium condition:

$$<\text{Step 1}> \quad = \quad <\text{Step 2}>$$

$$\sigma \times \pi\, d \quad = \quad P \times \frac{\pi}{4} d^2$$

$$\boxed{P \quad = \quad \frac{4\sigma}{d}}$$

Case 2

Surface tension on a hollow bubbles (Soap bubbles)

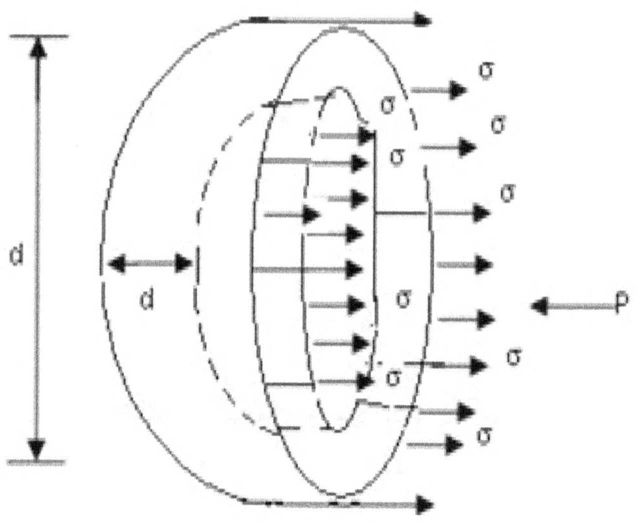

Figure 1.2: Hollow Bubble

Step 1

Tensile force acting on a surface= $\quad 2\lfloor \sigma \times \pi d \rfloor \qquad$ (N)

Step 2

Pressure force acting on a surface= $\quad P \times \dfrac{\pi}{4} d^2 \qquad$ (N)

Step 3

For equilibrium condition:

<Step 1> $\quad = \quad$ <Step 2>

$$2\lfloor \sigma \times \pi d \rfloor \quad = \quad P \times \dfrac{\pi}{4} d^2$$

$$\boxed{P \quad = \quad \dfrac{8\sigma}{d}}$$

Case 3

Surface tension on a liquid jet,

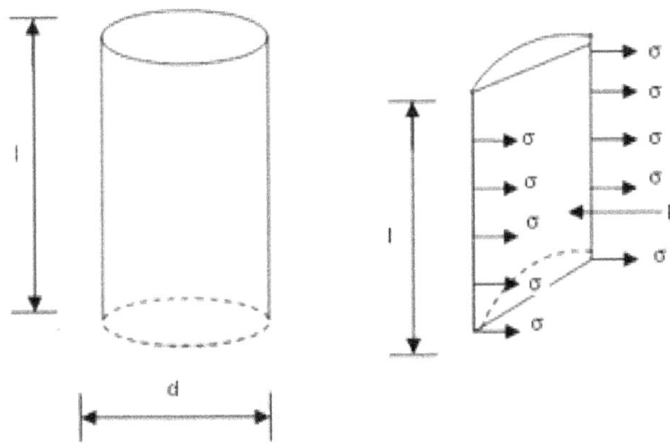

Figure 1.3: Liquid Jet

Step 1

Force due to pressure $= P \times d \times l$ (N)

Step 2

Force due to surface tension $= 2 \times \sigma \times l$ (N)

Step 3

For equilibrium condition

$$\text{<Step 1>} \quad = \quad \text{<Step 2>}$$

$$P \times d \times l \quad = \quad 2 \times \sigma \times l$$

$$P \quad = \quad \frac{2\sigma}{d}$$

Surface tension   -   Relationship between P, σ and d [J D S]

Liquid Jet	Liquid Droplet	Soap bubble or Hollow bubble
$P = \dfrac{2\sigma}{d}$	$P = \dfrac{4\sigma}{d}$	$P = \dfrac{8\sigma}{d}$

11) Capillarity

It is defined as phenomenon of rise or fall of liquid surface in a small tube relative to adjacent general level of liquid when the tube is held vertically in the liquid. The rise of liquid surface is known as capillarity rise while the fall of liquid surface is known as capillarity fall.

Derive an expression for capillarity rise.

Step 1

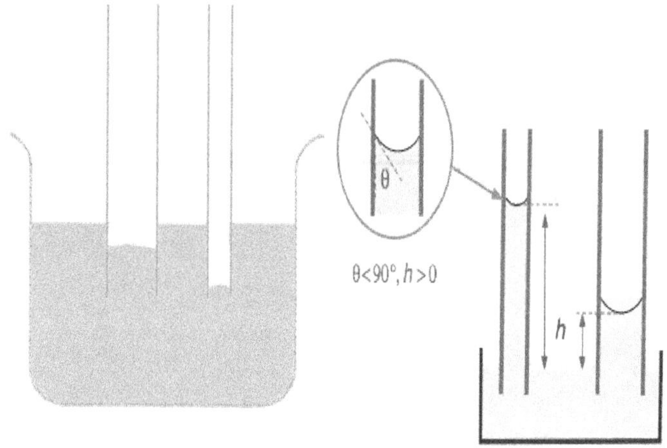

Figure 1.4: Capillarity Raise and Fall

Step 2

Where,

    σ -   Surface tension of the liquid

    θ -   Angle of contact between liquid and glass tube.

The weight of the liquid in the tube = Area of the tube x ρ x h x g (N)

$$= \frac{\pi}{4}d^2 \, x\, \rho \, x\, g \, x\, h \quad N$$

$$= m^2 \, x\, \frac{kg}{m^3} \, x\, \frac{m}{s^2} \, xm \quad N$$

Step 3

Vertical component of surface tension $= \sigma \; x \; \pi d \; x \; Cos\, \theta$ N

Step 4

For equilibrium condition

$$\langle \text{Step 1} \rangle = \langle \text{Step 2} \rangle$$

$$\frac{\pi}{4} d^2 \times \rho \times g \times h = \sigma \times \pi d \times Cos\ \theta$$

$$h = \frac{\sigma \times \pi d \times Cos\ \theta \ 4}{\pi d^2 \times \rho \times g}$$

$$h = \frac{4\sigma \times Cos\ \theta}{d.\rho.g}$$

where,

θ -   Angle of contact between liquid and glass tube

σ -   Surface Tension (N/m)

ρ -   Density (kg/m³)

g -   Gravity (m/s²)

d -   Diameter of the tube (m)

h -   Capillarity rise

Note

h = Positive value, then it is capillarity rise.

h = Negative value, then it is capillarity fall

For water $\theta = 0$ => $Cos\ \theta = 1$;

For mercury, $\theta = 128°$

12) Vapour Pressure

When the liquid is kept in a closed vessel, it evaporates into vapour and this vapour occupies the space between the free surface of the liquid and top of the vessel.

This vapour exerts a partial pressure on the free surface of the liquid. This pressure is known as vapour pressure P.

13) Cohesion

This is due to the force of attraction between molecules of same liquid.

14) Adhesion

It is defined as the force of attraction between the molecules of two different liquid.

Common Units

1 atmospheric pressure	=	1 Bar

$$\text{Pressure (P)} = \frac{Force}{Area} = \frac{N}{m^2}$$

1 atm pressure	=	1 bar	=	1×10^5 N/m²
1 Cent	=	435.67 sq.ft.	
100 cents	=	1 acre	

Young's modulus	Pressure (P)	Stress (σ)	Bulk modulus (K)
$E = \dfrac{Stress}{Strain} = \dfrac{N}{m^2}$	$P = \dfrac{F}{A} = \dfrac{N}{m^2}$	$S = \dfrac{L}{A} = \dfrac{N}{m^2}$	$K = \dfrac{dP}{\frac{dV}{V}} = \dfrac{N}{m^2}$

Equivalent Unit

Force	=	1 N	=	1 kg.m/s²
		1 N.m	=	1 J
		1 J/s	=	1 W

Compare the Units

Acceleration	Kinematic Viscosity
a (or) g = m/s²	γ = m²/s

Table 1.1: Basic Dimensions

Basic dimensions			Same unit
M	L	T	Work or energy =
mass	Length	Time	W = F x distance W = Nm
Kg	M	S	Torque T = F x perpendicular distance = Nm
Force (F) = m x a = kg x m/s² 1 N = MLT⁻²			

Table 1.2: Dimensions and Units

Sl. No.	Description	Symbol and Formula	Dimension
1	Area	$A = L \times L = m^2$	L^2
2	Volume	$V = L \times L \times L = m^3$	L^3
3	Velocity	$v = \dfrac{Dis\tan ce}{Time} = \dfrac{m}{s}$	$\dfrac{L}{T} = LT^{-1}$
4	Angular Velocity	$w = \dfrac{v}{r} = \dfrac{m}{sxm} = \dfrac{1}{s}$	$\dfrac{1}{T} = T^{-1}$
5	Acceleration	$a\ (or)\ g = \dfrac{Velocity}{Time} = \dfrac{m}{sxs} = \dfrac{m}{s^2}$	$\dfrac{L}{T^2} = LT^{-2}$
6	Angular Acceleration	$\propto = \dfrac{a}{r} = \dfrac{m}{s^2 xm} = \dfrac{1}{s^2}$	$\dfrac{1}{T^2} = T^{-2}$
7	Young Modulus (E), Pressure (P), Shear Stress (τ), Bulk Modulus (K)	$P\ or\ \tau\ or\ K = \dfrac{Force}{Area} = \dfrac{N}{m^2}$	$\dfrac{MLT^{-2}}{L^2} = ML^{-1}T^{-2}$
8	Density or Mass density	$\rho = \dfrac{Mass}{Volume} = \dfrac{kg}{m^3}$	ML^{-3}
9	Specific weight or weight density	$w = \dfrac{Weight\ or\ Force}{Volume} = \dfrac{N}{m^3}$	$\dfrac{MLT^{-2}}{L^3} = ML^{-2}T^{-2}$
10	Viscosity or dynamic viscosity	$\mu = \dfrac{NS}{m^2}$	$\dfrac{MLT^{-2}.T}{L^2} = ML^{-1}T^{-}$
11	Kinematic Viscosity	$\gamma = \dfrac{m^2}{s}$	$\dfrac{L^2}{T} = L^2T^{-1}$
12	Surface tension	$\sigma = \dfrac{N}{m}$	$\dfrac{MLT^{-2}}{L} = MT^{-2}$
13	Work or Torque (T) Energy (W)	$W = F \times Distance = N\text{-}m$	$MLT^{-2}.L = ML^2T^{-2}$
14	Power	$P = \dfrac{Workdone}{Time} = \dfrac{N.m}{s}$	$\dfrac{MLT^{-2}.L}{T} = ML^2T^{-}$
15	Momentum	$P = m \times v = kg.\ m/s$	$\dfrac{ML}{T} = MLT^{-1}$
16	Discharge	$Q = \dfrac{Quantity\ of\ water\ delivered}{Time} = \dfrac{m^3}{s}$	$\dfrac{L^3}{T} = L^3T^{-1}$

1.3. Newton Law of Viscosity

Newton law of viscosity: $\tau \, \alpha \, \dfrac{du}{dy}$

Where,

τ -   Shear stress.

$\dfrac{du}{dy}$ -   Shear strain.

u   -   Dynamic Viscosity.

$$\tau \; = \; u.\dfrac{du}{dy}$$

1.4. Types of Fluids

1) Ideal fluid
2) Real fluid
3) Newtonion fluid
4) Non-newtonion fluid
5) Ideal Plastic fluid.

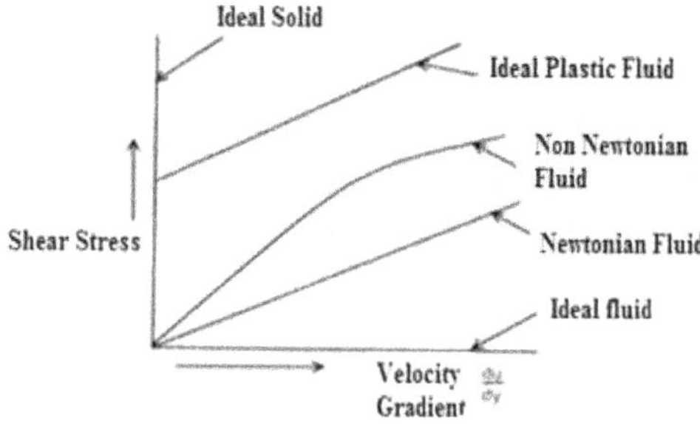

Figure 1.5: Types of Fluids

1) Ideal Fluid

The fluid which is incompressible (ρ=c) and is having no viscosity is called ideal fluid.

2) Real Fluid

The fluid, which is compressible ($\rho \neq$C) and is having viscosity is called real fluid.

Eg. Gas. All the fluid in actual practice are real fluid.

3) Newtonion Fluid

The fluid in which the shear stress is directly proportional to Shear strain is known as Newtonion Fluid (obeys the Newton law)

$$\tau \;=\; u.\frac{du}{dy}$$

Where

τ -   Shear stress;

$\dfrac{du}{dy}$ -   Shear strain or velocity gradient (eg. Water)

u   -   Dynamic Viscosity.

4) Non-Newtonion Fluid

The fluid in which the shear stress is not directly proportional to shear strain is known as Non-newtonion fluid (does not obey Newton law).

(eg. Blood, Paste)

5) Ideal Plastic Fluid

The fluid in which shear stress is more than the yield value and shear stress is directly proportional to shear strain is called ideal plastic fluid. (Obeys the newton law)

$$\tau \;=\; u.\frac{du}{dy}$$

Table 1.3: Gas laws

Boyle's law	Charle's law	Gaws Law / Joutes law	Renault's law	Avagardo's law
$v\alpha\dfrac{1}{P}$ PV = C	$v\alpha T$ $\dfrac{v}{T}=C$ (P = C)	$U \propto T$	R = Cp – Cv	Mv = Constant μ - Molecular weight of gas γ - Specific volume of gas
It states that volume of a given mass of a gas varies inversely as its absolute pressure when temperature remain constant	It states that the volume of a given mass of a gas varies directly as its absolute temperature when pressure remains constant	It states that the internal energy of a given quantity of gas depends only on the temperature	It states that the two specific heat Cp and Cv of a gas do not change with the change of temperature and pressure	It states that equal volume of different perfect gases at the same temperature and pressure contain equal number of molecules.

Problems in Fluid Properties

1. Calculate the specific weight, density and specific gravity of one litre of a liquid which weighs 1 N.

Given

Volume (v)   =   1 lit.

         =   1/1000 m³

Weight (W) =   1 N

To Calculate

ω, ρ, s =   ?

Step 1

(i) Specific weight, ω = $\dfrac{weight}{volume} = \dfrac{7}{\dfrac{1}{1000}}$ N/m³

ω =   7000 N/m³

Step 2

(ii) Density, ρ

ω = $\rho \times g$

ρ = $\dfrac{w}{g} = \dfrac{7000}{9.8} \dfrac{N/m^3}{m/s^2}$

ρ =   713.56 kg/m³

Step 3

(iii) Specific gravity, s

S = $\dfrac{\rho_{liquid}}{\rho_{std.liquid}} = \dfrac{713.56}{1000}$

S =   0.713

2. Velocity distributed over a plate is given by u = $\dfrac{2}{3}y - y^2$ in which v is the v is the velocity

in m/s have a distance y meter above the plate, determine the shear stress at y = 0 and y = 0.15 m. Take dynamic viscosity of fluid as 8.63 poise.

Given

$$u = \dfrac{2}{3}y - y^2 \quad (u = 8.63 \text{ poise} = 0.863 \text{ NS/m}^2$$

u → Velocity in m/s

y → Distance in 'm'

Determine

Shear Stress τ $\quad = \quad$?

$\quad$ At y $\quad = \quad$ 0

$\quad$ At y $\quad = \quad$ 0.15 m

Step 1

Shear stress τ $\quad = \quad \mu.\dfrac{du}{dy}$

$\quad$ u $\quad = \quad \dfrac{2}{3}y - y^2$

$\quad \dfrac{du}{dy} \quad = \quad \dfrac{2}{3} - 2y$

$\quad \left(\dfrac{du}{dy}\right)_{y=0} \quad = \quad 2/3$

$\quad \left(\dfrac{du}{dy}\right)_{y=0.15} \quad = \quad 0.37$

Shear stress at y = 0,

$$\tau = 0.863 \,(2/3)$$
$$\tau_{y=0} = 0.575 \text{ N/m}^2$$

Shear stress at y = 0.15τ $\quad = 0.863 \,(0.37)$

$$\tau_{y=0.15} = 0.319 \text{ N/m}^2$$

3. A Newtonian fluid is filled in the clearance between a shaft and a concentric sleeve. The sleeve attains a speed of 50 cm/s when a force of 40 N is applied to the sleeve parallel shaft. Determine the speed with a force of 200 N applied.

Given

$$F_1 = 40 \text{ N} \qquad\qquad u_1 = 50 \text{ cm/s} \qquad 0.5 \text{ m/s}$$

$$F_2 = 200 \text{ N} \qquad\qquad u_2 = ?$$

$$T = \mu.\frac{du}{dy} \qquad\qquad \frac{F}{A} = \mu.\frac{du}{dy}$$

$$\frac{F}{u} = \mu.\frac{A}{dy}$$

Step 1   ***Step 2***

$$\frac{F_1}{u_1} = \mu.\frac{A}{dy} \;\dots\; (1) \qquad\qquad \frac{F_2}{u_2} = \mu.\frac{A}{dy} \;\dots (2)$$

$$(1) = (2)$$

$$\frac{F_1}{u_1} = \frac{F_2}{u_2}$$

$$\frac{40}{0.5} = \frac{200}{u_2}$$

$$u_2 = \frac{200 x 0.5}{40}$$

$u_2 = 2.5$ m/s => Speed when F = 200 N

4. A plate 0.025 mm distance from a fixed plate, moves at 60 cm/s and requires a force of 2 N per unit area to maintain this speed. Determine the fluid viscosity between the plates?

Given

$dy = 0.025$ mm $= 0.025$ x 10^{-3} m

$du = 60$ cm/s $= 0.6$ m/s

$F = 2\text{N};$

$A = 1$ m^2

Determine

Viscosity, $\mu = ?$

Solution

Step 1

$$T = \mu \frac{du}{dy}$$

Step 2

$$\frac{F}{A} = \mu \frac{du}{dy}$$

$$\frac{2}{1} = \mu \frac{0.6}{0.025x10^{-3}}$$

$$\mu = \frac{2x0.025x10^{-3}}{0.6} = 0.083 \times 10^{-3} \text{ N.S/m}^2$$

5. Determine the intensity of shear of an oil having viscosity = 1 poise. The oil is used for lubricating the clearance between a shaft of diameter 10 cm and its journal bearing the clearance is 1.5 mm and the shaft rotate at 150 rpm.

   NOTE: Relationship between U.D.N

   $$U = \frac{\pi DN}{60} \text{ m/s.}$$

Given

μ	=	1 poise =	$\frac{1}{10}$ N.S/m^2	=	0.1 N.S/m^2
d	=	10 cm =	0.1 m		
dy	=	1.5 mm =	0.0015 m		
N	=	150 rpm			
T	=	?			

Solution

Step 1

$$\tau = \mu \frac{du}{dy} = 0.1 \text{ x } \frac{du}{1.5x10^{-3}}$$

Step 2

$$Du \quad = \quad U \quad = \quad \frac{\pi DN}{60} = \frac{\pi x0.1x150}{60} = 0.79$$

$$\tau \quad = \quad \frac{0.1 x 0.79}{1.5 x 10^{-3}} \quad = \quad 0.0523 \times 10^3 \quad = \quad 52.3 \text{ N /m}^2$$

6. Calculate dynamic viscosity of an oil which is used for lubrication between the square plate of size 0.8 x 0.8 m and an inclined plane with angle of inclination 30° as shown in figure. The weight of the square plate is 300 N and it slides down the inclined plane with a uniform velocity of 0.3 m/s. The thickness of oil film is 1.5 mm.

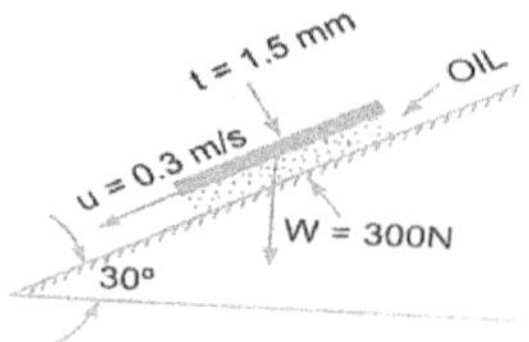

Figure 1.6

Given

A = 0.8 x 0.8 = 0.64 m²

U = 0.3 m/s

dy = 1.5 mm = 1.5 x 10⁻³ m

W = 300 N

θ = 30°

Determine

$$\mu \quad = \quad ?$$

Solution

Step 1

$$\tau \quad = \quad \mu \frac{du}{dy}$$

Step 2

$$\frac{F}{A} \quad = \quad \mu \frac{du}{dy}$$

Step 3

$$f = W \text{ Cos } \theta = 300 \text{ Cos } 30° = 259.81 \text{ N}$$

$$\frac{259.81}{0.64} = \mu \frac{0.3}{1.5 x 10^{-3}}$$

Step 4

$$\mu = \frac{259.81x1.5x10^{-3}}{0.64x0.3} = 2.02 \text{ N.s/m}^2$$

7. A 15 cm diameter vertical cylinder rotates concentrically inside another cylinder of diameter 15.10 cm. Both cylinders are 25 cm high, the space between the cylinders is filled with a liquid whose viscosity is unknown. With a torque 12 N.m is required to rotate the inner cylinder at 100 rpm. Determine the viscosity of the fluid.

Given

D_i	=	15 cm	=	0.15 m
D_o	=	15.10 cm		0.1510 m
T	=	12 N.m		
N	=	100 rpm (inside)		

Determine

$$\mu = ?$$

Solution

Step 1

$$\text{Shear stress, } \tau = \mu\frac{du}{dy} = \frac{F}{A}$$

Step 2

Area	=	πdl m^2 =	π x 15 x 25	=	1177.5 cm^2= 0.118 m^2
T	=	F x radius (i)			
12	=	F x 7.5 cm			
F	=	$\dfrac{12}{0.075}$ =	160 N		

Step 3

$$T = \frac{160}{0.118} = 1355.9$$

Step 4

$$U = \frac{\pi D_1 N}{60} = \frac{\pi x 0.15 x 100}{60} = 0.785 \text{ m/s}$$

Step 5

$$Dy = \frac{D_o - D_i}{2} = \frac{0.151 - 0.15}{2} = \frac{0.001}{2} = 0.0005 \text{ m}$$

$$T = \mu\frac{du}{dy}$$

$$\mu = \tau\frac{du}{dy}$$

$$= \frac{1355.9 \times 0.0005}{0.785} = 0.86 \text{ N-s/m}^2$$

8. An oil of viscosity 5 poise is used for lubrication between a shaft and sleeve. The diameter of the shaft is 0.5 m and it rotates at 200 rpm. Calculate power lost in oil for a sleeve length 100 mm. The thickness of oil film is 1 mm.

Given

μ	=	5 poise =	0.5 N.s/m^2
D	=	0.5 m	
N	=	200 rpm	
l	=	100 mm	= 0.1 m
dy	=	1 mm	= 1 c 10^{-3} m

Determine

Power = ?

Solution

Step 1

$$Power = \frac{2\pi NT}{60}$$

Step 2

$$T = F \text{ x radius}$$

Step 3

$$A = \pi dl = 3.14 \times 0.5 \times 0.10 = 0.157 \text{ m}^2$$

Step 4

$$T = \mu\frac{du}{dy}$$

Step 5

$$U \quad = \quad \frac{\pi DN}{60} \quad = \quad \frac{3.14 x 0.5 x 200}{60} \quad = \quad 5.2 \text{ m/s}$$

Step 6

$$\frac{F}{A} \quad = \quad \mu\frac{du}{dy}$$

$$\frac{F}{0.157} \quad = \quad \frac{5.2}{1x10^{-3}}\,x0.5$$

$$F \quad = \quad 5.2 \text{ x } 0.5 \text{ x } 0.157 \text{ x } 10^3 \quad = \quad 408.2 \text{ N}$$

Step 7

$$T \quad = \quad 408.2 \text{ x } \frac{0.5}{2} \quad = \quad 102.05 \text{ Nm}$$

Step 8

$$\text{Power} \quad = \quad \frac{2\pi NT}{60} = \frac{2x3.14x200x102.05}{60} = 2136.2 \text{ W} = 2.14 \text{ kW}$$

9. A dynamic viscosity of an oil used for lubrication between a shaft and sleeve is 6 poise. A shaft is of diameter 0.4 m and rotates at 190 rpm. Calculate power lost in the bearing for a sleeve length of 90 mm. The thickness of the film is 1.5 mm.

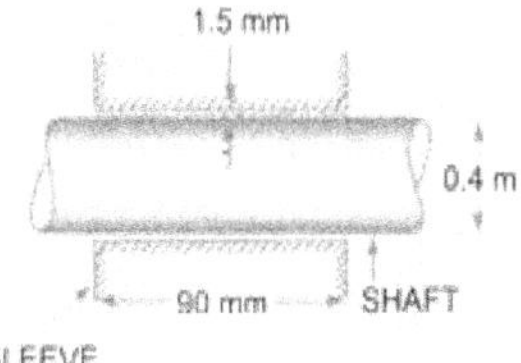

Figure 1.7

Given

μ	=	6 poise =	0.6 Ns/m²
D	=	0.4 m	
N	=	190 rpm	
P	=	?	
L	=	90 x 10⁻³ m	
Dy	=	1.5 mm = 1.5 x 10⁻³ m	

Solution

Step 1

$$\text{Power (P)} = \frac{2\pi NT}{60}$$

Step 2

$$T = \text{Force x radius}$$

Step 3

$$T = \mu\frac{du}{dy}$$

Step 4

$$U = \frac{\pi DN}{60} = \frac{3.14x0.4x190}{60} = 3.97 \text{ m/s}$$

Step 5

$$\frac{F}{A} = \mu\frac{du}{dy}$$

$$A = \pi Dl = 3.14 \times 0.4 \times 90 \times 10^{-3} = 0.113 \text{ m}^2$$

$$\frac{F}{0.113} = 0.6\frac{3.97}{1.5x10^{-3}}$$

$$F = 0.6\frac{3.97}{1.5x10^{-3}}x0.113 = 179.4 \text{ N}$$

Step 6

$$T = \text{Force x D/2} = 179.4 \times 0.2 = 35.88 \text{ Nm}$$

Step 7

$$\text{Power} = \frac{2\pi NT}{60} = \frac{2x3.14x190x35.88}{60}$$

$$= 713.53 \text{ W}$$

$$= 0.714 \text{ kW}$$

10. Two large plane surfaces are 2.4 cm apart. The space between the surface is filled with glycerin. What force is required to drag a very thin plate of surface area 0.5 m² between the two large plane surfaces a speed of 0.6 m/s. If –

(i) A thin plate is in the middle of the two plane surfaces and

(ii) A thin plate at a distance of 0.8 cm from one of the plane surface?

Take dynamic viscosity of glycerin is equal to 8.1 x 10⁻¹ N.s/m²

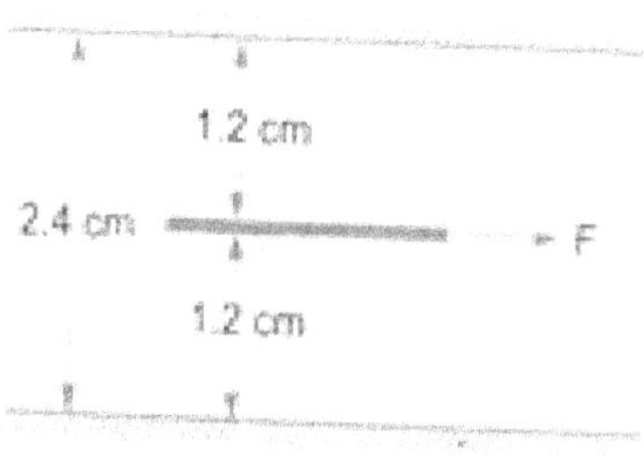

Figure 1.8

Given

A	=	0.5 m²
U	=	0.6 m/s
μ	=	8.1 x 10⁻¹ Ns/m²
F	=	?

Solution

Step 1

Case (i)

$$F = F_1 + F_2$$

$$\frac{F_1}{A} = \mu \frac{du}{dy} = 8.10 x 10^{-1} x \frac{0.6}{0.012}$$

$$F_1 = 40.5 \text{ N x } 0.5 = 20.25 \text{ N}$$

$$\frac{F_2}{A} = \mu \frac{du}{dy} = 8.10 x 10^{-1} x \frac{0.6}{0.012}$$

$$F_2 = 40.5 \text{ N x } 0.5 = 20.25 \text{ N}$$

$$F = F_1 + F_2$$

$$F = 40.5 \text{ N}$$

Step 2

Case (ii)

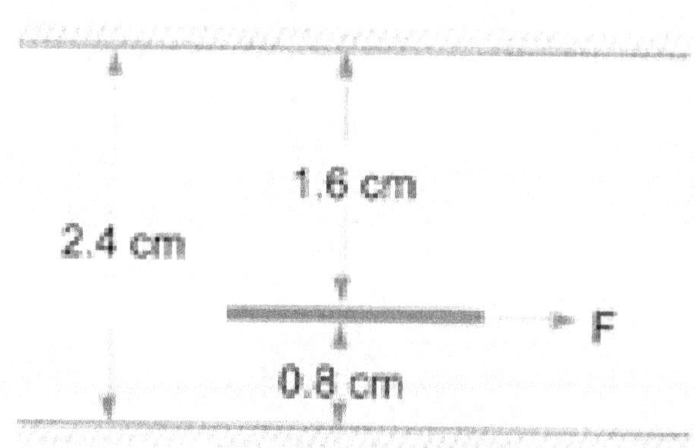

Figure 1.9

F	=	$F_1 + F_2$
$\dfrac{F_1}{A} = \mu\dfrac{du}{dy}$	=	$8.10 \times 10^{-1} \times \dfrac{0.6}{0.016} \times 0.5$
F_1	=	20.25 N
$\dfrac{F_2}{A} = \mu\dfrac{du}{dy}$	=	$8.10 \times 10^{-1} \times \dfrac{0.6}{0.008} \times 0.5$
F_2	=	30.37 N
F	=	$F_1 + F_2$
F	=	45.56 N

11. If the velocity profile of a fluid over a plate is a parabolic with the vertex 20 cm from the plate, where the velocity is 120 cm/s. Calculate the velocity gradient and shear stresses at a distance of 0, 10 and 20 cm from the plane, if the viscosity of the fluid is 8.5 poise.

$\dfrac{du}{dy}$	=	?
T	=	0
y	=	0, 10, 20 cm
μ	=	8.5 poise = 0.85 N.s/m²

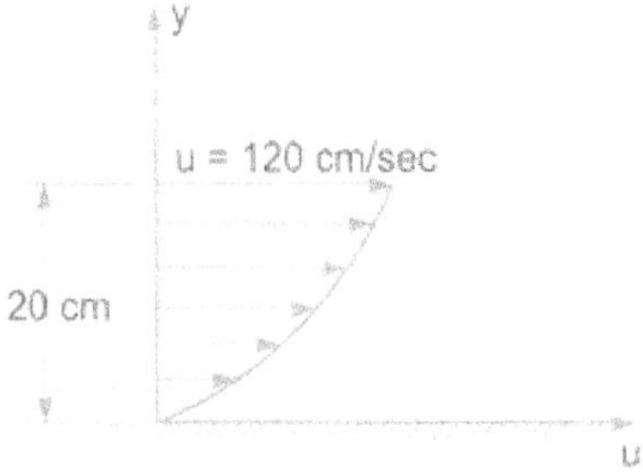

Figure 1.10

Step 1

The velocity profile is given parabolic equation

$$u = ay^2 + by + c \qquad ..... (i)$$

(a, b, c are constant)

Step 2

Case (i)

Boundary condition

$$y = 0; \ u = 0$$

Step 3

Case (ii)

$$y = 20 \text{ cm} = 0.2 \text{ m}$$
$$u = 120 \text{ cm/s} = 1.2 \text{ m/s}$$

Step 4

Case (iii)

$$y = 0.2 \text{ m} \qquad \frac{du}{dy} = 0$$

Case (i) $y = 0; u = 0$

   (1)   => $0 = C$

Case (ii) $y = 0.2 \text{ m}; \ u = 1.2 \text{ m/s}$

   (1)   => $1.2 = a\,(0.04) + 0.2\,b$

   $$0.04\,a + 0.2\,b = 1.2 \qquad ..... (ii)$$

Case (iii) $y = 0.2$ m; $\dfrac{du}{dy} = 0$

$$\dfrac{du}{dy} = 2ay + b$$

$$2ay + b = 0$$

$$0.4\,a + b = 0 \qquad\qquad \text{... (iii)}$$

Step 5

Solve (ii) and (iii) equation

$$0.04\,a + 0.2\,b = 1.2$$

$$0.4\,a + b = 0$$

(ii) x 1 $0.04\,a + 0.2\,b = 1.2$

                  $-\qquad\quad -\qquad\qquad -$

(iii) x 0.1 $0.04\,a + 0.1\,b = 0$

$$0.1\,b = 1.2$$

$$b = 12$$

equation (iii) =>

$$0.4\,a + 12 = 0$$

$$0.4\,a = -12$$

$$a = -30$$

Step 6

Substituting a, b, c values in equation (i)

$$u = ay^2 + by + c$$

$$= -30\,y^2 + 12\,y$$

$$\dfrac{du}{dy} = -60\,y + 12$$

Step 7

Velocity Gradient

$$\dfrac{du}{dy} = -60\,y + 12$$

$$\frac{du}{dy}_{(y=0)} \quad = \quad 12 \ 1/s$$

$$\frac{du}{dy}_{(y=0.1m)} \quad = -60\,(0.1) + 12 \quad = \quad 6 \ \ 1/s$$

$$\frac{du}{dy}_{(y=0.2m)} \quad = -60\,(0.2) + 12 \quad = \quad 0$$

Step 8

Shear stress:

$$\tau \quad = \quad \mu\frac{du}{dy}$$

$\tau_{(y=0)}$	$=$	$0.85 \times (-60\,(0) + 12)$
$\tau_{(y=0)}$	$=$	$10.2 \ N/m^2$
$\tau_{(y=0.1)}$	$=$	$0.85 \times (-60\,(0.1) + 12)$
$\tau_{(y=0.1)}$	$=$	$5.1 \ N/m^2$
$\tau_{(y=0.2m)}$	$=$	$0.85 \times (-60\,(0.2) + 12)$
$\tau_{(y=0.2m)}$	$=$	$0 \ N/m^2$

12. A Surface tension of water in contact with air at 20°C is 0.0725 N/m. The pressure inside a droplet of water is to be 0.02 N/cm² greater than the outside pressure. Calculate the diameter of the droplet of water.

Given

$$\sigma \quad = \quad 0.0725 \ N/m$$
$$P \quad = \quad 0.02 \ N/cm^2 = 0.02 \times 10^4 = 200 \ N/m^2$$
$$T \quad = \quad 20°C$$
$$d \quad = \quad ?$$

Solution

$$\text{Liquid droplet} \quad P \quad = \quad \frac{4\sigma}{d}$$

$$d \quad = \quad \frac{4\sigma}{P} = \frac{4 \times 0.0725}{200}$$

$$= \quad 1.45 \times 10^{-3} m$$

13. Problem in Capillarity

Calculate the capillarity rise in a glass tube of 2.5 mm diameter when immersed vertically in

1. Water

2. Mercury

Take surface tension σ = 0.0725 N/m for water and σ = 0.52 N/m for Mercury in contact with air. The specific gravity for mercury is given as 13.6 and angle of contact is 130°.

Given

$$d \quad = \quad 2.5 \text{ mm} = \quad 2.5 \times 10^{-3} \text{ m}$$
$$\sigma_{water} \quad = \quad 0.0725 \text{ N/m}$$
$$\sigma_{mercury} \quad = \quad 0.52 \text{ N/m}$$
$$S_{water} \quad = \quad 1$$
$$S_{mercury} \quad = \quad 13.6$$
$$\theta_{Hg} \quad = \quad 130; \quad \theta_{water} \quad = \quad 0 \text{ (known)}$$

Solution

Step 1

Case (i) water.

$$h \quad = \quad \frac{4\sigma Cos\ \theta}{d\rho g} = \frac{4 x 0.0725 x Cos\ 0}{2.5 x 10^{-3} x 1000 x 9.81} = 0.0118 \text{ m Capillarity rise}$$

Step 2

Case (ii) Mercury

$$h \quad = \quad \frac{4\sigma Cos\ \theta}{d\rho g} = \frac{4 x 0.52 x Cos\ 130}{2.5 x 10^{-3} x 13600 x 9.81} \quad = \quad -4 x 10^{-3} \text{ m}$$

$$h_{hg} \quad = \quad -0.004 \text{ m} \qquad \text{Capillarity fall}$$

1.5. Kinematics of Flow

It is defined as the brand of science deals with motion of particle without considering the force.

1.5.1. Types of Fluid Flow

1. Steady and unsteady flow.

2. Uniform and Non-uniform flow.

3. Laminar and Turbulent flow.

4. Compressible and Incompressible flow.

5. Rotational and Irrotational flow.

6. One, two and three dimensional flow.

Steady Flow	Unsteady flow
It is defined as a type of flow in which fluid characteristics like velocity, pressure, and density at a point do not change with time. $$\frac{\partial v}{dt} = c; \frac{\partial P}{dt} = c; \frac{\partial \rho}{dt} = c$$	It is defined as a type of flow in which fluid characteristics like velocity, pressure and density at a point change with time. $$\frac{\partial v}{dt} \neq c; \frac{\partial P}{dt} \neq c; \frac{\partial \rho}{dt} \neq c$$
Uniform Flow	Non-uniform flow
It is defined as the type of flow in which the velocity at any given time do not change with respect to the space (i.e. the length of direction of flow) $$\frac{\partial v}{ds} = c$$	It is defined as the type of flow in which the velocity at any given time change with respect to the space (i.e. the length of direction of flow) $$\frac{\partial v}{ds} \neq c$$
Laminar flow	Turbulent flow
It is defined as a type of flow in which the fluid particle move well defined path or stream line (all the stream are straight and parallel) or Reynolds number Re<2000	It is defined as the tube of flow in which the fluid particle move in zig-zag manner (or) Reynolds number Re>4000
Compressible flow	Incompressible flow
It is defined as the flow in which the density of the fluid changes from point $$\rho \neq c \quad \text{eg. Gas}$$	It is defined as the flow in which the density is constant for fluid flow. $$\rho = c \quad \text{eg. Water}$$
Rotational flow	Irrotational flow
It is defined as the fluid particle while flowing along stream line also rotate about its own axis. Eg. Turbine	The fluid particle while flowing along the stream line do not rotate about the own axis.

1D flow	2 D flow	3 D flow
It is that of flow in which the flow parameter such as velocity is a function of time and one space coordinate only say x.	It is that of flow in which the flow such as velocity is a function of time and two rectangular space coordinate say x, y	It is that of flow in which the velocity is a function of time and three mutually perpendicular direction say x, y, z.
$u = f(x)$ $v = 0$ $\omega = 0$	$u = f_1(x, y)$ $v = f_2(x, y)$ $\omega = 0$	$u = f_1(x, y, z)$ $v = f_2(x, y, z)$ $\omega = f_3(x, y, z)$

1.6. Rate of Flow or Discharge (Q)

It is defined as quantity of water delivered per second (m^3/s)

$$Q = A \times v \quad = \quad m^3/s$$

Where, $A \quad = \quad$ Area in m^2, $\quad v \quad =$ Velocity in m/s

1.7. Continuity Equation

$$Q_1 = \rho_1 A_1 V_1$$

$$Q_2 = \rho_2 A_2 V_2$$

$$Q = \rho_1 A_1 V_1 = \rho_2 A_2 V_2$$

$$Q = A_1 V_1 = A_2 V_2$$

$(\rho_1 = \rho_2)$ ----------- Incompressible flow

$$Q = \rho_1 A_1 V_1 = \rho_2 A_2 V_2$$

$(\rho_1 \neq \rho_2)$ ----------- Compressible flow

1.8. Derive the Expression for Continuity Equation in three Dimension

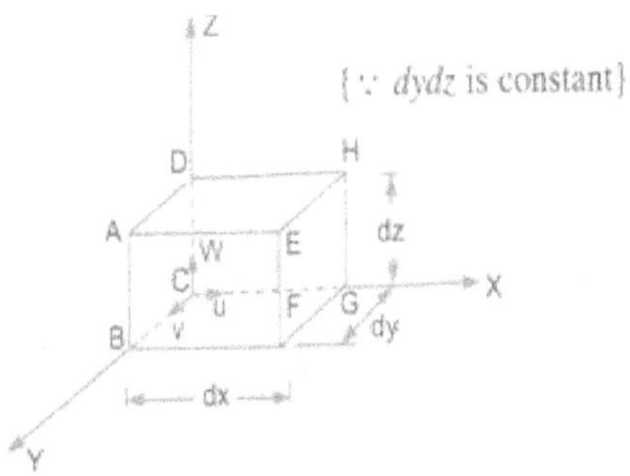

Figure 1.11

Step 1

Consider a fluid element of length dx, dy, dz in the direction x, y and z. Let u, v, ω are the initial velocity component in x, y and z direction as shown in figure 1.6.

Step 2

Mass of fluid entry face.

ABCD per second　　　　=　　　　ρ x velocity (in x-direction) x Area of ABCD

　　　　　　　　　　　=　　　　ρ x u x dy x dz

Step 3

Mass of fluid leaving a face.

EFGH per second　　　　=　　　　$\rho \times u \times dy \times dz \quad + \frac{\partial}{\partial x}\left[\rho \ x \ u \ x \ dy \ x \ dz\right]dx$

Step 4

Gain of mass in x direction

　　　　　　　=　　　　Entering　　　　　　　　　　Leaving

　　　　　　　　　　　Mass in ABCD　　　　　　　　Mass in EFGH

　　　　　　　=　　　　$\rho \times u \times dy \times dz - \rho \times u \times dy \times dz - \frac{\partial}{\partial x}\left[\rho \ x \ u \ x \ dy \ x \ dz\right]dx$

Gain of mass in x direction　　　　=　　　　$-\frac{\partial}{\partial x}\left[\rho \ x \ u \ x \ dy \ x \ dz\right]dx$

　　　　　　　　　　　　　　　=　　　　$-\frac{\partial}{\partial x}\left[\left(\rho \ x \ u \ \right)x \ dx \ x \ dy \ x \ dz\right]$

Step 5

Similarly

Mass of gain in y direction $\quad$ = $\quad -\dfrac{\partial}{\partial y}\left[\rho \ x \ v \ x \ dx \ x \ dz\right]dy$

Mass of gain in y direction $\quad$ = $\quad -\dfrac{\partial}{\partial y}\left[(\rho \ x \ v \ )x \ dx \ x \ dy \ x \ dz\right]$

Step 6

Mass of gain in z direction $\quad$ = $\quad -\dfrac{\partial}{\partial z}\left[\rho \ x \ \omega \ x \ dx \ x \ dy\right]dz$

Mass of gain in z direction $\quad$ = $\quad -\dfrac{\partial}{\partial y}\left[(\rho \ x \ \omega \ )x \ dx \ x \ dy \ x \ dz\right]$

Step 7

Net gain of mass = $\quad -\left[\dfrac{\partial}{\partial x}(\rho x u)+\dfrac{\partial}{\partial y}(\rho x v)+\dfrac{\partial}{\partial z}(\rho x \omega)\right]dx \ x \ dy \ x \ dz$

The net increase of mass per unit time in the fluid element must be equal to rate of increase of mass in the fluid element.

$$-\left[\frac{\partial}{\partial x}(\rho x u)+\frac{\partial}{\partial y}(\rho x v)+\frac{\partial}{\partial z}(\rho x \omega)\right]dx \ x \ dy \ x \ dz = \frac{\partial \rho}{\partial t}dx \ x \ dy \ x \ dz$$

$$-\left[\frac{\partial}{\partial x}(\rho x u)+\frac{\partial}{\partial y}(\rho x v)+\frac{\partial}{\partial z}(\rho x \omega)\right] = \frac{\partial \rho}{\partial t}$$

Condition 1:

Flow is steady $\dfrac{\partial \rho}{\partial t}$ $\qquad$ = $\qquad$ 0

Condition 2:

$\qquad$ Flow is incompressible ρ $\qquad$ = $\qquad$ constant

$$-\rho\left[\frac{\partial u}{\partial x}+\frac{\partial v}{\partial y}+\frac{\partial \omega}{\partial z}\right] \qquad = \qquad 0$$

$$\left[\frac{\partial u}{\partial x}+\frac{\partial v}{\partial y}+\frac{\partial \omega}{\partial z}\right] \qquad = \qquad 0 \qquad \text{(3 dimensional)}$$

$$\left[\frac{\partial u}{\partial x}+\frac{\partial v}{\partial y}\right] \qquad = \qquad 0 \qquad \text{(2 dimensional)}$$

1.9. Dynamics of Flow

It is the study of fluid motion with forces causing the flow.

State Bernoulli's theorem for steady flow of incompressible fluid. Derive an expression for Bernoulli's theorem from first principle and state the assumption made for such a derivation.

(OR)

What is Euler's equation of the motion? How will you obtain Bernoulli's equation from it.

Step 1

Assumptions

1. Fluid is ideal (viscosity is zero)
2. Fluid flow is steady (dt = 0)
3. Fluid is incompressible (ρ = C)
4. The flow is irrotational.

Step 2

Statement of Bernoulli's Theorem

The total energy at any point of the fluid is constant. The total energy consists of pressure energy, Kinetic energy and Potential energy.

OR

$$\frac{P}{\rho g} + \frac{v^2}{2g} + z \quad = \quad 0$$

Where, $\dfrac{P}{\rho g}$ = Pressure head

$\dfrac{v^2}{2g}$ = Kinetic head

z = Potential head / Datum head.

Step 3

Euler's Equation of Motion

This equation of motion in which the forces due to gravity and pressure are taken into consideration.

Consider a stream line in which flow taking place in 'S' direction as shown in figure.

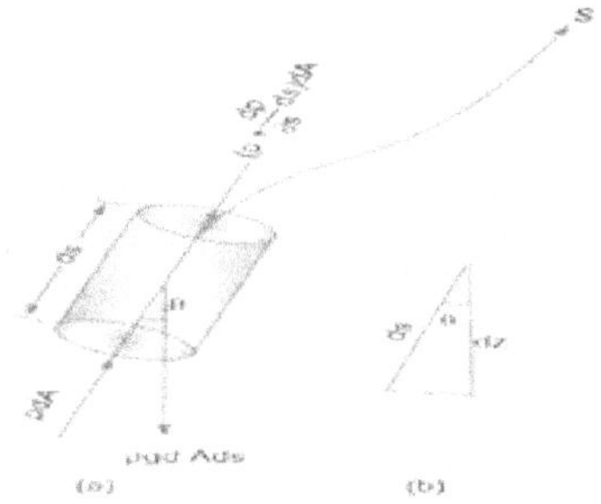

Figure 1.12: Forces on a Fluid Element

Step 4

Consider a cylindrical element of cross section dA and distance dS.

Step 5

Force acting on the cylindrical elements are

1. Pressure force p.dA in the direction of flow (+)

2. Pressure force $\left[P + \dfrac{\partial P}{\partial s} ds \right] dA$ in the opposite direction of flow (-ve)

3. Weight of element "$\rho gdA.ds.Cos\ \theta$" acting downward (-ve)

Let 0 be the angle between a direction of flow and line of action of weight element.

Therefore, Resultant force

$$PdA - \left[P + \frac{\partial P}{\partial s} ds \right] dA - \rho gdA.ds.Cos\ \theta = \text{Resultant force.}$$

Step 6

$$
\begin{aligned}
F &= & m \times a \\
&= & m \times a_s \\
F &= & \rho\, dA.ds. \times a_s
\end{aligned}
$$

Step 5 = Step 6

$$PdA - \left[P + \frac{\partial P}{\partial s} ds \right] dA - \rho gdA.ds.Cos\ \theta = \rho\, dA.ds. \times a_s$$

$$PdA - PdA - \frac{\partial P}{\partial s} ds.dA - \rho gdA.ds.Cos\ \theta = \rho\, dA.ds. \times a_s$$

$$-\frac{\partial P}{\partial s}ds.dA - \rho g dA.ds.Cos\,\theta \;\; = \rho\,dA.ds.x\,a_s \qquad\ldots\ldots\ldots\ldots \qquad (i)$$

$$a_s \quad = \quad ?$$

$$a_s \quad = \quad \frac{Velocity}{time} = \frac{dv}{dt}$$

Applying Chain Rule

$$a_s \quad = \quad \frac{\partial V}{\partial s}.\frac{ds}{dt} + \frac{\partial v}{\partial t}$$

Substitute the Value of as in (1)

$$(i)\;\rightarrow\quad -\frac{\partial P}{\partial s}ds.dA - \rho g dA.ds.Cos\,\theta - \rho\,dA.ds.\left[\frac{\partial V}{\partial s}.\frac{ds}{dt} + \frac{\partial v}{\partial t}\right] = 0$$

$$V = \frac{ds}{dt}\;;\text{ Flow is steady } \frac{dv}{dt} = 0$$

$$(i)\;=>\quad -\frac{\partial P}{\partial s}ds.dA - \rho g dA.ds.Cos\,\theta - \rho\,dA.ds.\;v\frac{\partial v}{\partial s} \quad = 0$$

$$\frac{\partial P}{\partial s}ds.dA + \rho g dA.ds.Cos\,\theta + \rho\,dA.ds.\;v\frac{\partial v}{\partial s} \quad = 0$$

Divided by PdA.ds on both Sides

$$(i)\;=>\quad \frac{\partial P}{\partial s}\frac{1}{\rho} + gCos\,\theta + v\frac{\partial v}{\partial s} = 0$$

$$\frac{\partial P}{\partial s}\frac{1}{\rho} + g\frac{dz}{ds} + v\frac{\partial v}{\partial s} \quad = 0$$

Multiply ∂s on both Sides

$$\frac{\partial P}{\rho} + gdz + v\partial v \quad = 0$$

Divided g by both Sides

$$\frac{\partial P}{\rho g} + gdz + \frac{v\partial v}{g} \quad = 0$$

Step 7

Both side integral

$$\int \frac{\partial p}{\rho} + \int gdz + \int V\partial v = \int 0$$

$$\frac{P}{\rho} + gz + \frac{V^2}{2} = C$$

$\% \; g$

$$\frac{P}{\rho g} + \frac{\cancel{g}z}{\cancel{g}} + \frac{V^2}{2g} = C$$

$$\frac{P}{\rho g} + Z + \frac{V^2}{2g} = C$$

This equation is called as Bernoullis equation.

1.10. Difference between Orifice meter and Venturimeter

Orifice	Venturimeter
	convergent · throat · Divergent
Circular plate inside the hole.\n\nFlow measuring devices.\n\nCd = 0.64	Flow measuring devices\n\nCd = 0.98

1.11. Give Practical Application of Bernoulli's Equation

1. Venturimeter
2. Orifice meter
3. Pitot Tube

Problems in Continuity Equation

1. The diameter of a pipe at the section 1 and 2 are 10 cm and 15 cm. Find the discharge through the pipe the velocity of water flowing through the pipe at section 1 is 5 m/s. Determine also velocity at section 2.

Given

$\qquad d_1 \qquad = \qquad 10 \; cm = 0.1 \; m$

$\qquad d_2 \qquad = \qquad 15 \; cm = 0.15 \; m$

$\qquad v_1 \qquad = \qquad 5 \; m/s.$

To find: v_2 and discharge (Q)

Solution

Step 1

Discharge Q $=$ $A_1 V_1$ $=$ $A_2 V_2$

$$= \frac{\pi}{4} d_1^2 x v_1 = \frac{\pi}{4} x 0.1^2 x 5 = \qquad 0.04 \text{ m}^3/\text{s}$$

$$0.04 = \frac{\pi}{4} d_2^2 x v_2 = \frac{\pi}{4} x 0.15^2 x v_2$$

Step 2

$$V_2 = \frac{4 x 0.04}{\pi x (0.15)^2} = 2.2 \text{ m/s}$$

2. A 30 cm diameter pipe conveying water branches into 2 pipes of diameter 20 cm and 15 cm respectively. If the average velocity in the 30 cm diameter pipe is 2.5 m/s. Find the discharge in this pipe, also determine velocity in 15 cm pipe if the average velocity in 20 cm diameter pipe is 2 m/s.

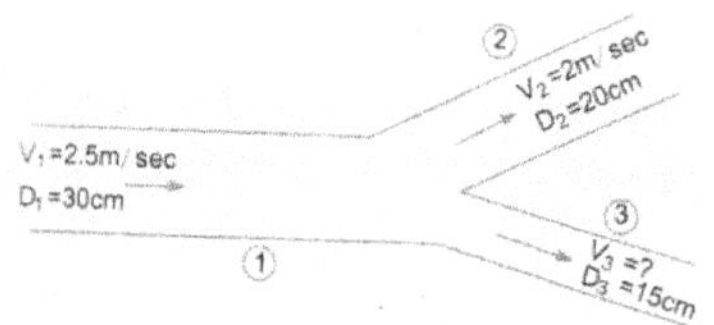

Figure 1.13

Given

d_1	=	30 cm	=	0.3 m
V_1	=	2.5 m/s.		
d_2	=	20 cm	=	0.2 m
V_2	=	2 m/s		
d_3	=	15 cm	=	0.15 m
V_3	=	?	$Q =$	?

Solution

Step 1

$$Q_1 = Q_2 + Q_3$$

Step 2

$$Q_1 = A_1 V_1 = \frac{\pi}{4}(0.3)^2 x 2.5 = 0.176 \text{ m}^3/\text{s}$$

Step 3

$$Q_2 = A_2 V_2$$

$$= \frac{\pi}{4}(0.2)^2 \, x2 = 0.063 \quad \text{m}^3/\text{s}$$

Step 4

$$Q_1 = Q_2 + Q_3$$

Step 5

$$Q_3 = Q_1 - Q_2 = 0.176 - 0.063 = 0.113 \ \text{m}^3/\text{s}$$

$$Q_3 = A_3 V_3$$

Step 6

$$V_3 = \frac{Q_3}{A_3} = \frac{0.113}{\frac{\pi}{4}(0.15)^2} = \frac{0.113x4}{3.14x(0.15)^2}$$

$$= 6.39 \ \text{m/s}$$

3. Water flows through a pipe AB 1.2 m diameter at 3 m/s and then passes through a pipe BC 1.5 m diameter. At C, the pipe branches. Branch CD is 0.8 m in diameter carries one third of the flow in AB. The flow velocity in CE is 2.5 m/s. Find the volume rate of flow in AB, the velocity in BC, the velocity in CD and diameter of CE

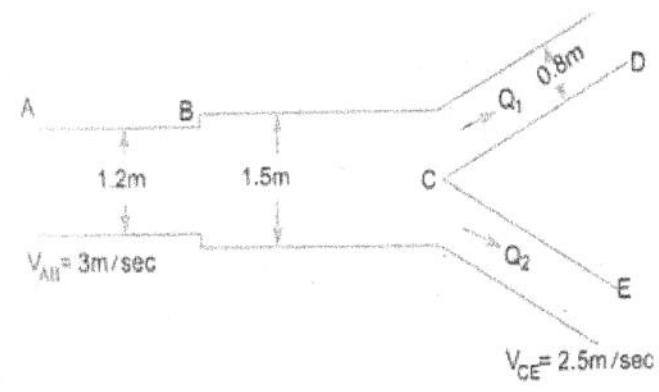

Figure 1.14

Given

d_{AB}	=	1.2 m		
V_{AB}	=	3 m/s.	Q_{AB} = ?	
d_{BC}	=	1.5 m	V_{BC} = ?	
d_{CD}	=	0.8 m	Rate of flow CD = 1/3 in AB	
V_{CD}	=	?		
d_{CE}	=	?		
V_{CE}	=	2.5 m/s		

Step 1

$$Q_{AB} = A_{AB} \times V_{AB}$$

$$= \frac{\pi}{4}(1.2)^2 x3 = 3.39 \text{ m}^3/\text{s}$$

Step 2

$$Q = Q_1 + Q_2$$

Step 3

$$Q_{CD} = 1/3 \ Q_{AB} = (1/3) \times 3.39 = 1.13 \text{ m}^3/\text{s}$$

$$Q_{CD} = A_{CD} \times V_{CD}$$

$$= \frac{\pi}{4}d_{CD}^{\ 2}xV_{CD}$$

Step 4

$$V_{CD} = \frac{1.13}{\frac{\pi}{4}0.8^2} = \frac{1.13x4}{\pi x0.8^2} = 2.256 \text{ m/s}$$

Step 5

$$Q_{AB} = Q_{CD} + Q_{CE}$$

$$3.39 - 1.13 = Q_{CE}$$

Step 6

$$Q_{CE} = 2.26 \text{ m}^3/\text{s}$$

$$Q_{CE} = A_{CE} \times V_{CE}$$

$$Q_{CE} = \frac{\pi}{4}d_{CE}^{\ 2}xV_{CE}$$

$$d^2_{CE} = \frac{Q_{CE}}{V_{CE}}\frac{4}{\pi} = \frac{2.26x4}{2.5x3.14}$$

$$= 1.17 \text{ m}$$

Step 7

$$Q_{BC} = V_{BC} \times A_{BC}$$

$$V_{BC} = \frac{Q_{BC}}{A_{BC}} = \frac{3.39}{\frac{\pi}{4}(1.5)^2} = \frac{3.39x4}{3.14x1.5^2}$$

$$= 1.9 \text{ m/s}$$

Problems in Bernoulli's Equation

1. A water is flowing through a pipe having diameter 20 cm and 10 cm at section 1 and 2. The rate of flow through the pipe 35 l/s. The section 1 is 6 m above datum and section 2 is 4 m above datum; if the pressure at section 1 is 39.24 N/cm². Find the intensity of pressure at section 2.

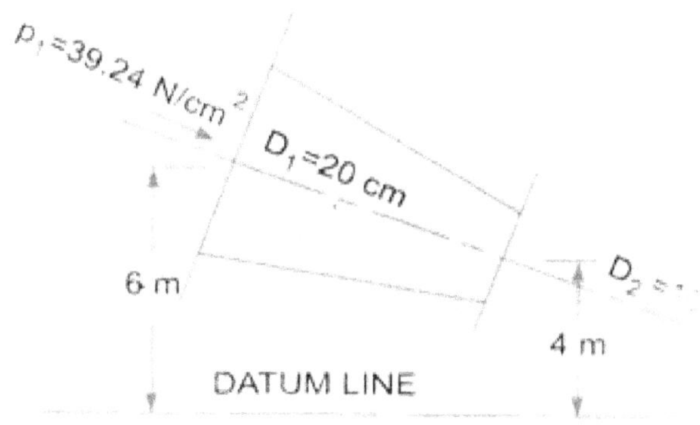

Figure 1.15

Given

d_1 = 20 cm = 0.2 m

d_2 = 10 cm = 0.1 m

Q = 35 lit/s = 0.0035 m³/s

P_1 = 39.24 N/cm² = 39.24 x 10⁴ N/m²

Step 1

By applying Bernoulli's equation

$$\frac{P_1}{\rho g} + \frac{v_1^2}{2g} + z_1 = \frac{P_2}{\rho g} + \frac{v_2^2}{2g} + z_2 \qquad \text{....} \qquad (1)$$

Step 2

Q = $A_1 V_1$

Step 3

$$V_1 = \frac{Q}{A_1} = \frac{35 \times 10^{-3}}{\frac{\pi}{4} d_1^2} = 3$$

$$= \frac{0.035x4}{\pi x(0.2)^2}$$

$$= 1.115 \text{ m/s}$$

Step 4

$$Q = A_2 \times V_2$$

$$V_2 = \frac{Q}{A_2} = \frac{0.035x4}{\pi(0.1)^2} = 4.459 \qquad \text{m/s}$$

(1) =>

$$\frac{39.24x10^4}{1000x9.81} + \frac{(1.115)^2}{2x9.81} + 6 = \frac{P_2}{1000x9.81} + \frac{(4.459)^2}{2x9.81} + 4$$

$$40 + 0.063 + 6 \quad = \quad \frac{P_2}{9810} + 1.013 + 4$$

Step 5

$$P_2 = 41.05 \times 9810 = 402700.5 = 40.3 \times 10^4 \text{ N/m}^2$$

2. Water is flowing through a tapper pipe of length 100 m having diameters 600 mm at the upper end and 300 mm at the lower end at the rate of 50 lit/sec. The pipe has a slope of 1 in 30. Find the pressure at the lower end if the pressure at the higher level is 19.62 N/cm²

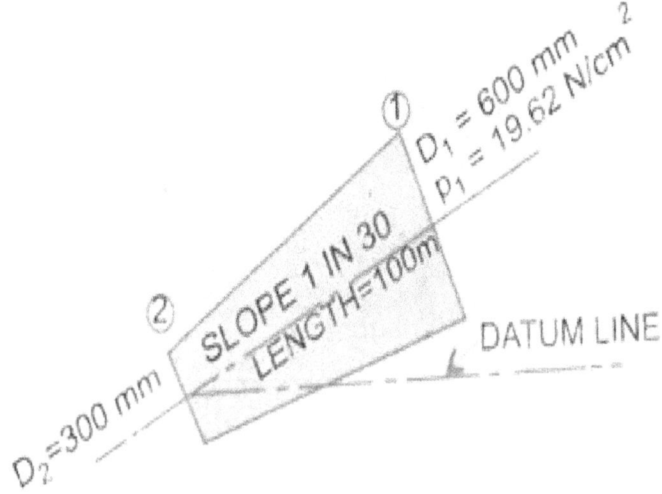

Figure 1.16

Given

Length (l)	=	100 m
d_1	=	600 mm = 0.6 m
d_2	=	300 mm = 0.3 m
Q	=	50 lit/sec.
	=	$\dfrac{50}{1000}$ m³/s
P_1	=	19.62 N/cm² = 19.62 x 10⁴ N/m²
P_2	=	?

Solution

Step 1

By applying Bernoullie's equation in section 1 and 2.

$$\frac{P_1}{\rho g} + \frac{v_1^2}{2g} + z_1 = \frac{P_2}{\rho g} + \frac{v_2^2}{2g} + z_2 \qquad\qquad (1)$$

Step 2

$$Q \quad = \quad A_1 V_1 \quad = \quad V_1 \quad = \frac{Q}{A_1} = \frac{0.05x4}{\pi x 0.6^2} \qquad 0.177 \text{ m/s}$$

Step 3

$$Q \quad = \quad A_2 V_2 \quad = \quad V_2 \quad = \frac{Q}{A_2} = \frac{0.05x4}{\pi x 0.3^2} \qquad 0.708 \text{ m/s}$$

$$(1) => \qquad \frac{19.62x10^4}{1000x9.81} + \frac{0.177^2}{2x9.81} + 3.33 = \frac{P_2}{1000x9.81} + \frac{0.708^2}{2x9.81} + 0$$

$$20 + 0.0015 + 3.33 = \frac{P_2}{9810} + 0.0255$$

$$23.306 = \frac{P_2}{9810}$$

Step 4

P_2 = 228631.8

OR

22.8 x 10⁴ N/m²

1.12. Required Formula for Venturimeter and Orifice Meter

Venturimeter	Orifice Meter
Actual Discharge (Q)	1. Actual discharge Q
$$Q_{act.} = Cdx\frac{a_1 x a_2}{\sqrt{a_1^2 - a_2^2}}x\sqrt{2gH}$$	$$Q_{act} = Cdx\frac{a_1 x a_2}{\sqrt{a_1^2 - a_2^2}}x\sqrt{2gH}$$
Cd = Coefficient of discharge = 0.98	Cd = Coefficient of discharge (0.64)
a_1 = Area at section (1) $\frac{\pi}{4}d_1^2$ m²	a_1 = Area at section (1) $\frac{\pi}{4}d_1^2$ m²
d_1 = Diameter of pipe	d_1 = Diameter of pipe
a_2 = Area at section (2) $\frac{\pi}{4}d_2^2$ m²	a_2 = Area at section (2) $\frac{\pi}{4}d_2^2$ m²
d_2 = Diameter of pipe (2) h = Pressure head (U tube manometer) NOTE:	d_2 = Diameter of pipe (2) h = Pressure head (U tube manometer) NOTE:
Case 1: $h = x\left[\dfrac{S_h}{S_l} - 1\right]$	**Case 1:** $h = x\left[\dfrac{S_h}{S_l} - 1\right]$
Let the differential manometer contains liquid which is heavier than the liquid following through a pipe X – Difference of heavier liquid column in U tube S_h – Specific gravity of heavier liquid S_l - Specific gravity of lighter liquid	Let the differential manometer contains liquid which is heavier than the liquid following through a pipe X – Difference of heavier liquid column in U tube S_h – Specific gravity of heavier liquid S_l - Specific gravity of lighter liquid
Case 2: If the differential manometer contain a liquid which lighter than the liquid flowing through the pipe $h = x\left[1 - \dfrac{S_l}{S_h}\right]$	**Case 2:** If the differential manometer contain a liquid which lighter than the liquid flowing through the pipe $h = x\left[1 - \dfrac{S_l}{S_h}\right]$

1. Pressure (P) is converted into pressure head (H)

$$h = \frac{P}{\rho g} \quad \{P = \rho gH\}$$

2. Differential head:

$$h = \frac{P_1}{\rho g} - \frac{P_2}{\rho g}$$

3. Coefficient of discharge

$$Cd = \sqrt{\frac{h - h_f}{h}}$$

4. Assume that 4% of differential head is lost between inlet and throat

$$h_f \quad = \quad \frac{4}{100}xh$$

5. Unit Conversion

   - The pressure at inlet is 17.658 N/cm²

   $P_1 = 17.658 \times 10^4 \text{ N/m}^2$

   - A vacuum pressure at the throat is 30 cm of mercury

$$\frac{P_2}{\rho g} = \text{ - 30 cm of Hg} = -\frac{30}{100} \text{ m of Hg}$$

$$\frac{P_2}{\rho g} = -\frac{30}{100}x13.6 \text{ m of water}$$

1.13. Defined Pitot Tube and Required Formula for Pitot Tube

Definition

The Pitot Tube is a device used for measuring the velocity of flow at any point in a pipe or channel.

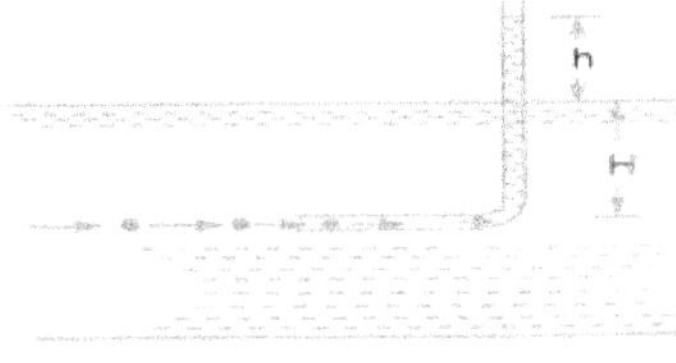

Figure 1.17: Pitot Tube

Required Formula

Central Velocity (v) or Velocity at center

$$v \quad = \quad Cv \times \sqrt{2gH}$$

where, Cv   =   Coefficient of velocity(0.98)

$$Q \quad = \quad A \; x \; \bar{v}$$

Where, $\bar{v}$ - Mean velocity

A – Area of the tube $\left(\dfrac{\pi}{4}d^2\right)$

Tips

- Static pressure in pipe is 100 mm of mercury (Vacuum)

Static pressure $\quad=\quad$ 100 mm of Hg. (Vacuum)

Static pressure head $\quad=\quad -\dfrac{100}{1000}$ m of Hg.

Static pressure head $\quad=\quad -\dfrac{100}{1000} x13.6$ m of water

- Stagnation pressure at the center of the pipe recorded by the pitot tube is 0.981 N/cm²

Stagnation pressure $\quad=\quad$ 0.981 N/Cm²

Static pressure head $\quad=\quad \dfrac{0.981x10^4}{\rho x g}$ m

Static pressure head $\quad=\quad \dfrac{0.981x10^4}{1000x9.81}$ m

$\qquad h \qquad = \qquad x\left[\dfrac{S_h}{S_l}-1\right]$

Problems in Venturimeter

1. A horizontal venturimeter with inlet and throat diameter 30 cm and 15 cm is used to measure the flow of water. The reading of differential manometer connected to the inlet and the throat is 20 cm of mercury. Determine the rate of flow Cd = 0.98

Given

d_1 $\quad=\quad$ 30 cm $\quad=\quad$ 0.3 m

d_2 $\quad=\quad$ 15 cm $\quad=\quad$ 0.15 m

x $\quad=\quad$ 20 cm of Hg. $\quad=\quad$ 0.2 m of Hg

Cd $\quad=\quad$ 0.98

Solution

Step 1

Actual Discharge (Q_{act})

$$Q_{act} \quad=\quad Cdx\dfrac{a_1 x a_2}{\sqrt{a_1^2-a_2^2}}\sqrt{2gh}$$

Step 2

$$h = x\left[\frac{S_h}{S_l} - 1\right]$$

$$= 0.2\left[\frac{13.6}{1} - 1\right] \qquad = \qquad 0.2\,(12.6) = 2.52 \text{ m of water}$$

Step 3

$$a_1 = \frac{\pi}{4}d_1^{\,2} = \frac{\pi}{4}(0.3)^2 \qquad = \qquad 7.065 \text{ m}^2$$

Step 4

$$a_2 = \frac{\pi}{4}d_2^{\,2} = \frac{\pi}{4}(0.15)^2 \qquad = \qquad 0.0177 \text{ m}^2$$

Step 5

$$Q_{act} = \frac{0.98x7.065x0.0177}{\sqrt{7.065^2 - 0.0177^2}}\,x\sqrt{2x9.81x2.52} = \frac{0.123}{7.06}x7.0127$$

$$Q_{act} = \qquad 0.123 \text{ m}^3/\text{s}$$

2. A horizontal venturimeter with inlet diameter 20 cm and throat diameter 10 cm is used to measure the flow of water. The pressure at inlet is 17.658 N/cm² and the vaccum pressure at the throat is 30 cm of mercury find the discharge of water through Venturimeter. Cd – 0.98

Given

$$d_1 \qquad = \qquad 20 \text{ cm} \quad = 0.2 \text{ m}$$

$$d_2 \qquad = \qquad 10 \text{ cm} \quad = 0.1 \text{ m}$$

$$P_1 \qquad = \qquad 17.658 \text{ N/cm}^2$$

$$\frac{P_1}{\rho g} \qquad = \qquad \frac{17.658x10^4}{9.81x1000} \quad \text{m}$$

$$\frac{P_1}{\rho g} \qquad = \qquad 18 \text{ m}$$

$$\frac{P_2}{\rho g} \qquad = \qquad 30 \text{ cm of Hg (vaccum)}$$

$$\frac{P_1}{\rho g} \qquad = \qquad -\frac{30}{100}x13.6 \qquad = -\,4.08 \text{ m of water}$$

Solution

Step 1

Actual Discharge (Q_{act})

$$Q_{act} = Cd x \frac{a_1 x a_2}{\sqrt{a_1^2 - a_2^2}} \sqrt{2gh}$$

Step 2

$$h = \frac{P_1}{\rho g} - \frac{P_2}{\rho g} \qquad = \qquad 18 + 4.08$$

$$h = 22.08 \text{ m}$$

Step 3

$$a_1 = \frac{\pi}{4} d_1^2 = \frac{\pi}{4}(0.2)^2 \qquad = \qquad 0.0314 \text{ m}^2$$

Step 4

$$a_2 = \frac{\pi}{4} d_2^2 = \frac{\pi}{4}(0.1)^2 \qquad = \qquad 0.0078 \text{ m}^2$$

Step 5

$$Q_{act} = \frac{0.98 x 0.0314 x 0.0078}{\sqrt{0.0314^2 - 0.0078^2}} x \sqrt{2 x 9.81 x 22.08} = \frac{0.00024}{0.0304} x 20.8$$

$$Q_{act} = 0.164 \text{ m}^3/\text{s}$$

3. The inlet and throat diameter of a horizontal venturimeter are 30 cm and 10 cm respectively. A liquid flowing through meter is water. The pressure intensity at inlet is 13.734 N/cm² while the vaccum pressure head at the throat 37 cm of mercury. Find the rate of flow. Assume that 4% of differential head is lost between the inlet and throat. Find the values of Cd for the venturimeter.

Given

Venturimeter

$$d_1 = 30 \text{ cm} = 0.3 \text{ m}$$

$$d_2 = 10 \text{ cm} = 0.1 \text{ m}$$

$$P_1 = 13.734 \text{ N/cm}^2$$

$$\frac{P_1}{\rho g} = \frac{13.734 x 10^4}{9.81 x 1000} \text{ m}$$

$$\frac{P_1}{\rho g} = \quad 14 \text{ m}$$

$$\frac{P_2}{\rho g} = \quad 37 \text{ cm of Hg (vaccum)} \quad = \quad \frac{-37}{100} x 13.6 = -5.03 \text{ m of water}$$

$$Q = \quad ?$$

$$Cd = \quad ?$$

Solution

Step 1

Actual Discharge (Q_{act})

$$Q_{act} = \quad Cdx\frac{a_1 x a_2}{\sqrt{a_1^2 - a_2^2}}\sqrt{2gh}$$

Step 2

$$h = \quad \frac{P_1}{\rho g} - \frac{P_2}{\rho g} \quad = \quad 14 + 5.03$$

$$h = \quad 19.03 \text{ m}$$

Step 3

$$a_1 = \quad \frac{\pi}{4}d_1^2 = \frac{\pi}{4}(0.3)^2 \quad = \quad 0.0707 \text{ m}^2$$

Step 4

$$a_2 = \quad \frac{\pi}{4}d_2^2 = \frac{\pi}{4}(0.1)^2 \quad = \quad 0.0078 \text{ m}^2$$

Step 5

$$h_f = \quad 4\% \text{ of } h$$

$$= \quad \frac{4}{100}x 19.03 \quad = \quad 0.761$$

Step 6

$$Cd = \quad \sqrt{\frac{h - h_f}{h}} = \sqrt{\frac{19.03 - 0.761}{19.03}} \quad = \quad 0.98$$

Step 7

$$Q_{act} = Cd \cdot \frac{a_1 x a_2}{\sqrt{a_1^2 - a_2^2}} \sqrt{2gh}$$

$$Q_{act} = \frac{9.8 x 0.0707 x 0.0078}{\sqrt{0.0707^2 - 0.0078^2}} x \sqrt{2 x 9.81 x 19.03} = \frac{0.00054}{0.0702} x 19.31$$

$$Q_{act} = 0.149 \text{ m}^3/\text{s}$$

Problems in Pitot Tube

1. A pitot tube is inserted in a pipe of 300 mm diameter. A static pressure in pipe is 100 mm of mercury (vaccum). The stagnation pressure at the center of pipe recorded by a pitot tube is 0.981 N/cm². Calculate the rate of flow of water if the mean velocity of water is 0.85 times of the central velocity. Take Cv = 0.98

Given

Pitot Tube

d	=	300 mm	=	0.3 m

$$\text{Static pressure} = 100 \text{ mm of Hg.(Vaccum)} = \frac{-100}{1000} x 13.6$$

Static pressure head = - 1.36

Cv = 0.98

Stagnation pressure = 0.981 N/cm²

$$= \frac{0.981 x 10^4}{1000 x 9.81} = 1 \text{ m}$$

Q = ?

$$\bar{v} = 0.85 \text{ x Central velocity}$$

Solution

Step 1

$$\text{Discharge (Q)} = A \times \bar{v}$$

Step 2

$$\bar{v} = 0.85 \text{ x v}$$

Step 3

$$v = Cv \times \sqrt{2gh}$$

Step 4

h	=	Stagnation Pressure head – Static pressure head
	=	1 – (-1.36)
	=	1 + 1.36
	=	2.36 m

Step 5

$$A = \frac{\pi}{4}d^2 = \frac{\pi}{4}(0.3)^2 = 0.0707 \text{ m}^2$$

Step 6

$$V = 0.98 \times \sqrt{2 x 9.81 x 2.36} = 6.66 \text{ m/s}$$

Step 7

$$\bar{v} = 0.85 \times 6.66 = 5.66 \text{ m/s}$$

Step 8

$$Q = A \times \bar{v} = 0.0707 \times 5.66 = 0.4 \text{ m}^3/\text{s}$$

Two Mark Questions-Basic Concepts and Properties

1. What is a real fluid?

   The fluid which is having the following properties is known as real fluids.

   a) It is compressible;

   b) They are viscous in nature

   c) Shear stress always exists in such fluids.

2. What are the properties of ideal fluids?

   Ideal fluids have following properties:

   a) It is not compressible.

   b) It has Zero viscosity.

   c) Shear force is Zero, when it is in motion.

3. Why are some fluid classified as Newtonian fluids? Give example to Newtonian fluids.

   In Newtonian fluids, a linear relationship between shear stress and rate of shear strain.

   Example: Water, Kerosene.

4. Define Density (or) Mass Density.

   $$\text{Density} = \rho = \frac{Mass}{Volume}\left(\frac{kg}{m^3}\right)$$

5. Define specific weight (or) Weight density

   $$\text{Specific weight} = W = \frac{Weight}{Volume}\left(\frac{N}{m^3}\right) \quad (OR) \quad W = \rho \times g \quad (N/m^3)$$

6. Define specific volume?

   $$\text{Specific volume} = \frac{Volume}{mass}\left(\frac{m^3}{kg}\right)$$

7. Define Specific gravity

   $$S = \frac{Specific\ Weight\ of\ Liquid}{Specific\ Weight\ of\ S\tan dard\ liquid\ Water}$$

8. Differentiate Compressibility and Bulk Modulus

Bulk Modulus	Compressibility
Symbol K	Symbol 1/K
K = Increase of Pressure / Volumetric strain	Volumetric Strain / Increase of Pressure
Unit = N/m²	Unit = m² / N

9. Define Viscosity of a fluid. (OR) Dynamics viscosity (or) N.

It is defined as the property of a fluid due to which it offers resistance to the movement of one layer of fluid over another adjacent layer.

$$\mu = \frac{\tau}{\dfrac{du}{dy}} = \frac{Shear\ Stress}{\dfrac{Change\ in\ Velocity}{Change\ in\ dis\tan ce}} \left(\frac{N-s}{m^2}\right)$$

10. Define Kinematic viscosity.

$$\gamma = \frac{Dynamic\ Vis\cos ity}{Density} \left(\frac{\mu}{\rho}\right) \qquad \left(\frac{m^2}{s}\right)$$

11. What is the effect of temperature on Viscosity of water and that of air?

When the temperature of water increase, the Viscosity will decreases.

When the temperature of air increases, the viscosity will increase.

12. Define Capillary.

Capillary is a phenomenon of rise or fall of liquid surface relative to the adjacent general level of liquid. This phenomenon is due to the combined effect of cohesion and adhesion of liquid particles. The rise of liquid level is known as capillary rise, whereas the fall of liquid surface is known as Capillary fall.

13. Explain the effect of property of capillary.

This phenomenon is due to the combined effect of cohesion and adhesion of liquid particle. So, the surface will act around the circumference of the tube.

14. Differentiate between Kinematic and Dynamic viscosity.

Dynamic Viscosity (or) Viscosity	Kinematic Viscosity
Shear stress α Shear Strain	γ = Dynamic Viscosity / Density
Symbol μ and Unit is Ns/m²	Symbol γ and Unit is m² / s

15. State the assumption made in deriving continuity equation
   a) The flow is incompressible
   b) The fluid is ideal

16. Suppose the small air bubbles in a glass of tap water may be on the order of 50 μ m in diameter. What is the pressure inside these bubbles?

Given: Dia of bubble, d　　　=　　　50 μ m = 50 x 10⁻⁶ m

Pressure difference inside the bubbles in excess of outside pressure

$$p = \frac{8\sigma}{d}$$

Assume σ = 0.0725 N/m

$$p = \frac{8x0.0725}{50x10^{-6}} = 11600\,N/m^2$$

Absolute pressure inside the air bubbles = 11600 + 101325 = 112925 N/m²

17. Why is it necessary in winter to use lighter oil for automobiles than in summer? To what property does the term lighter refer?

   - Lighter oil is good for easier cold weather starting and reducing friction.
   - The term lighter refers to viscosity of oils, which increases with the decrease in temperature.

18. Find the height of a mountain where the atmospheric pressure is 730 mm of Hg at Normal conditions.

   Let Atmospheric pressure at mean sea level,

   P_{atm} = 760 mm of Hg = 0.76 m of Hg = 0.76 x 13.6 x 9.81 = 101.396 kN/m² (or) 101396 N/m²

   Given   P   =   730 mm of Hg = 0.73 m of Hg

                       =   0.73 x 13.6 x 9.81

                       =   97.394 kN/m² or 97394 N/m²

               P   = ρ g h (h = P / ρ g)

   Height of mountain, $h = \dfrac{(P_{atm} - p)x1000}{w}$; where, w = ρ x g

$$h = \frac{(P_{atm} - p)x1000}{1.2x9.81} = 339.96\ m$$

19. What is Manometer?

   Manometer is a device, which is used for measuring the pressure and a point in a fluid by balancing the column of fluid by the same or another column of fluids.

20. Define Surface tension.

   Surface tension is due to the force of Cohesion between the liquid particles the free surface.

$$\sigma = \frac{Force}{Unit\ Length}\left(\frac{N}{m}\right)$$

21. Write the equation of surface tension of liquid jet, liquid droplet and soap bubble.

Liquid Jet	Liquid Droplet	Soap bubble
$P = \dfrac{2\sigma}{d}$	$P = \dfrac{4\sigma}{d}$	$P = \dfrac{8\sigma}{d}$

P = Pressure σ = Surface tension d = diameter

Dept. of Mechanical Engineering,
Fluid Mechanics and Machinery
III-Semester
Unit-I Introduction
Part-A

1. A soap bubble is formed when the inside pressure is 5 N/m^2 above the atmospheric pressure. If surface tension in the soap bubble is 0.0125 N/m, find the diameter of the bubble formed.

2. The converging pipe with inlet and outlet diameters of 200 mm and 150 mm carries the oil whose specific gravity is 0.8. The velocity of oil at the entry is 2.5 m/s, find the velocity at the exit of the pipe and oil flow rate in kg/sec.

3. What is the variation of viscosity with temperature for fluids?

4. Find the height of a mountain where the atmospheric pressure is 730 mm of Hg at Normal conditions.

5. What is meant by vapour pressure of a fluid?

6. Distinguish between atmospheric pressure and gauge pressure.

7. What are Non-Newtonian fluids? Give examples.

8. Mention the uses of a manometer.

9. What do you mean by absolute pressure and gauge pressure?

10. Define the term Kinematic Viscosity and give its dimension.

11. What is meant by continuum?

12. State Pascal's hydrostatic law.

13. What is specific gravity? How is it related to density?

14. How does the dynamic viscosity of liquids and gases vary with temperature?

15. How does the dynamic viscosity of (a) liquids and (b) gases vary with temperature?

16. What is the difference between gauge pressure and absolute pressure?

17. Differentiate between solids and liquids.

18. Define the following terms:

    a) Total pressure b) Centre (or) position of pressure.

19. What is meant by capillarity?

20. Define buoyancy.

21. What is viscosity? What is the cause of it in liquids and in gases?

22. State Pascal's law.

Part-B

1. A drainage pipe is tapered in a section running with full of water. The pipe diameters at the inlet and exit are 1000 mm and 500 mm respectively. The water surface is 2 m above the centre of the inlet and exit is 3 m above the free surface of the water. The pressure at the exit is 250 mm of Hg vacuum. The friction loss between the inlet and exit of the pipe is 1/10 of the velocity head at the exit. Determine the discharge through the pipe.

2. A pipe of 300 mm diameter inclined at 30° to the horizontal is carrying gasoline (specific gravity =0.82). A Venturi meter is fitted in the pipe to find out the flow rate whose throat diameter is 150 mm. The throat is 1.2 m from the entrance along its length. The pressure gauges fitted to the Venturi meter read 140 kN/m² and 80 kN/m² respectively. Find out the coefficient of discharge of Venturi meter if the flow is 0.20 m³/s.

3. Explain the properties of a hydraulic fluid.

4. 0.5 m shaft rotates in a sleeve under lubrication with viscosity 5 poise at 200 rpm. Calculate the power lost for a length of 100 mm if the thickness of the oil is 1 mm.

5. (i) Derive Bernoulli's theorem and state its limitations.

    (ii) A horizontal Venturi meter with inlet diameter 200 mm and throat diameter 100 mm is employed to measure the flow of water. The reading of the differential manometer connected to the inlet is 180 mm of mercury. If Cd = 0.98, determine the rate of flow.

6. Derive continuity equation from basic principles.

7. Derive Euler's equation of motion for flow along a stream line. What are the assumptions involved.

8. A horizontal pipe carrying water is gradually tapering. At one section the diameter is 150 mm and flow velocity is 1.5 m/s. If the drop in pressure is 1.104 bar at a reduced section, determine the diameter of that section. If the drop is 5 kN/m2, what will be the diameter - Neglect losses?

9. State Bernoulli theorem for steady flow of an incompressible fluid. Derive an expression for Bernoulli equation and state the assumptions made.

10. (i) A 15 cm diameter vertical pipe is connected to 10 cm diameter Vertical pipe with a reducing socket. The pipe carries a flow of 100 1/s. At point 1 in 15 cm pipe gauge pressure is 250 kPa. At point 2 in the 10 cm pipe located 1.0 m below point 1 the gauge pressure is 175 kPa.

    1. Find whether the flow is upwards / downwards.

    2. Head loss between the two points

    (ii) Differentiate Venturi meter and Orifice meter.

Unit II

Flow through Circular Conduits

2.1. Derive the Expression for Loss of Head Friction in Pipes (hf) OR Derive the Expression for Darcy-Weisbach Equation

Step 1

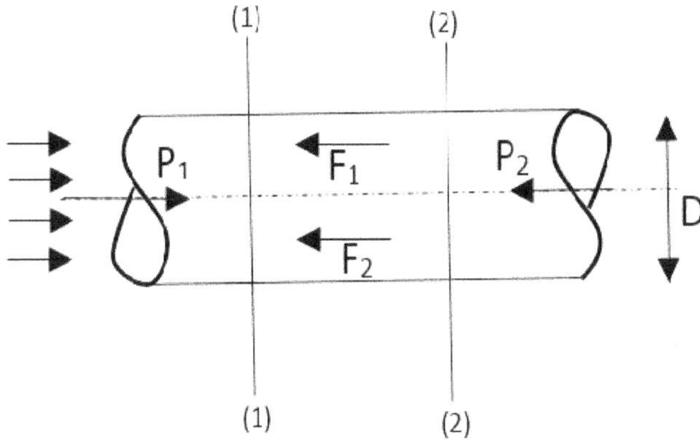

Step 2

Consider a uniform horizontal pipe having steady flow as shown in figure.

Let,

$$
\left.\begin{array}{l}
\text{1-1} \quad = \\
\text{2-2} \quad =
\end{array}\right\} \text{ two section of pipe}
$$

Let,

P_1 = Pressure Intensity at sec 1-1

P_2 = Pressure Intensity at sec 2-2

V_1 = velocity of flow

D = Diameter of pipe

L = Length of the pipe between sec 1-1 & 2-2

h_f = Loss of head due to friction

F' = Frictional resistance per unit wetted area per unit velocity.

Step 3

Applying Bernoulli's equation at sec 1-1 & 2-2

$$\frac{P_1}{\rho g} + \frac{V_1^2}{2g} + Z_1 = \frac{P_2}{\rho g} + \frac{V_2^2}{2g} + Z_2 + h_f$$

Horizontal Pipe $z_1 = z_2 =$ constant

$$\frac{P_1}{\rho g} + \frac{V_1^2}{2g} = \frac{P_2}{\rho g} + \frac{V_2^2}{2g} + h_f$$

Step 4

Diameter of section 1-1 and 2-2 are same

$V_1 = V_2$

$$\frac{P_1}{\rho g} = \frac{P_2}{\rho g} + h_f$$

$$h_f = \frac{P_1}{\rho g} - \frac{P_2}{\rho g} \quad .... \quad ..... \quad (1)$$

Step 5

Frictional force

F_1 = frictional resistance per unit wetted area per unit velocity * wetted area * V^2

$\quad = F' * \pi dl * V^2$

$P \quad = \pi d$

$F_1 \quad = F' * PL * V^2$

$F_1 \quad = F'PLV^2 ... \quad .... \quad (2)$

Step 6

Force acting on the fluid between sec 1-1 and sec 2-2

- Pressure force at section 1-1 $\quad = \quad P_1 A$
- Pressure force at section 2-2 $\quad = \quad -P_2 A$
- Frictional force $\quad = \quad -F_1$

Resolving above forces

$$P_1 A - P_2 A - F_1 = 0$$

$(P_1 - P_2) A = F_1$

$(P_1 - P_2) = F_1 / A$

$P_1 - P_2 = F'PlV^2 / A$

Step 7

From eq 1

$$\frac{P1 - P2}{\rho g} = h_f$$

$$P_1 - P_2 = h_f \cdot \rho g$$

Equate 2,3

$$h_f \cdot \rho g = \frac{F'Pl\,V^2}{A}, \; h_f = \frac{F'Pl\,V^2}{A\rho g}$$

We know that,

$$\frac{F'}{\rho} = \frac{F}{2}$$

$$\frac{P}{A} = \frac{\pi d}{\frac{\pi d^2}{4}} = \frac{4}{d}$$

$$h_f = \frac{f}{2} * \frac{4}{d} * \frac{lV^2}{g}$$

$$h_f = \frac{4flV^2}{2dg}$$

4f = Friction factor

This equation is called as Darcy-weisbach equation

2.2. Derive the Expression for Hagan Poise Willes Equation

Derive the expression for flow of viscous fluid through circular conduit pipe

Step 1

For the flow of viscous fluid through circular pipe the velocity distortion across the section, the ratio of maximum velocity to average velocity shear stress distribution and drop of pressure for a given length to be determined

Step 2

- The flow through a circular pipe will be laminar.
- If the Reynold's number is less than 2000.

Reynold's number

$$R_e = \frac{\mu D}{V}$$

U = velocity m/sec

D = Diameter m

V = kinematic viscosity m²/sec

Step 3

$$P = \frac{P.F}{A}$$

$$P.F = P.A$$

$$P.F = P.\,\pi r^2$$

Step 4

Consider a horizontal pipe of radius (R) the viscous fluid flowed from left to right in the pipe as shown in figure.

r = radius of the fluid element

Step 5

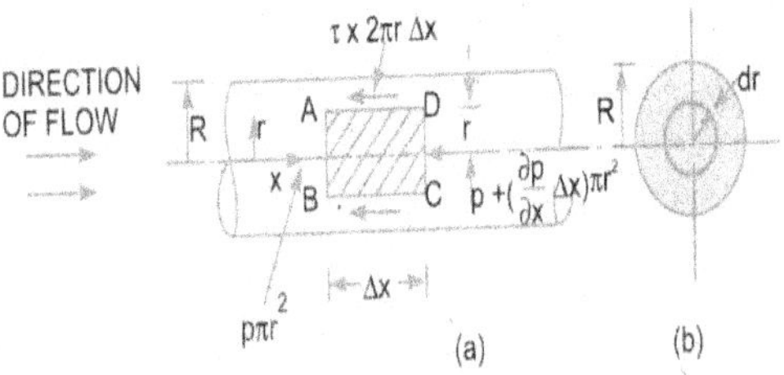

Figure 2.1: Viscous Flow through a Pipe

Step 6

The force acting on the fluid element are,

- Pressure force = Pπr² on face AB

- Pressure force = $-\left(P + \frac{\partial p}{\partial x}\Delta x\right)\pi r^2$ on the face CD

- Shear force = $-(\text{ᴛ} * 2\pi r * \Delta x)$

Therefore resolving above three force,

$$P\pi r^2 - \left(P + \frac{\partial p}{\partial x}\Delta x\right)\pi r^2 - (\text{ᴛ} * 2\pi r * \Delta x) = 0$$

$$P\pi r^2 - P\pi r^2 + \frac{\partial p}{\partial x}\Delta x\pi r^2 - \text{ᴛ} * 2\pi r * \Delta x = 0$$

$$-\frac{\partial p}{\partial x}\Delta x\pi r^2 = \text{ᴛ}2\pi r\Delta x$$

$$\text{ᴛ} = -\frac{r}{2}\frac{\partial p}{\partial x} \qquad \text{....} \qquad (1)$$

Step 7

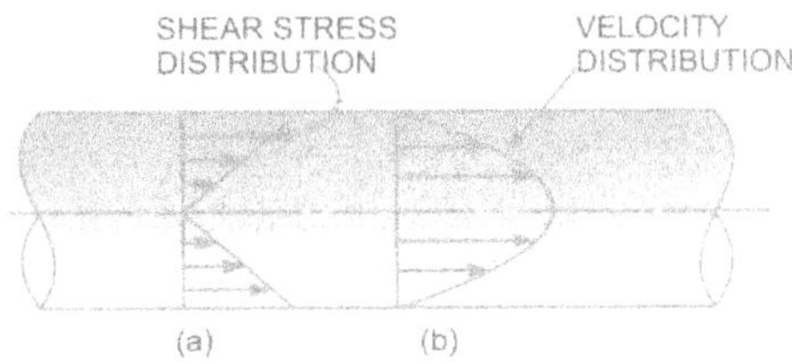

Figure 2.2: Shear Stress and Velocity Distribution Across a Section

Step 8

Newton's law of viscosity,

$$\text{T} = \mu \frac{du}{dy}$$

Y is measured form the pipe wall

$$y = R - r$$

Differentiate with respect to 'r'

$$dy = 0 - dr$$

$$dy = -dr$$

$$\text{T} = \mu \frac{du}{dy}$$

$$-\frac{r}{2}\frac{\partial p}{\partial X} = \mu \frac{du}{-dr}$$

$$du = \frac{r}{2\mu}\frac{\partial p}{\partial X}\, dr$$

Step 9

$$\int du = \int \frac{r}{2\mu}\frac{\partial p}{\partial X}\, dr$$

$$\int du = \frac{1}{2\mu}\frac{\partial p}{\partial X}\int r dr$$

$$u = \frac{1}{2\mu}\frac{\partial p}{\partial X}\frac{r^2}{2} + c$$

$$u = \frac{1}{4\mu}\frac{\partial p}{\partial X} r^2 + c \qquad \ldots \qquad \ldots \qquad \cdot\cdot \qquad (2)$$

Step 10

Where c is constant, c is obtained from boundary condition,

U = 0

r = R sub in equation (2)

$$0 = \frac{1}{4\mu} \frac{\partial p}{\partial X} R^2 + c$$

$$c = \frac{1}{4\mu} \frac{\partial p}{\partial X} R^2$$

Therefore C value in eq (2)

$$u = \frac{1}{4\mu} \frac{\partial p}{\partial X} r^2 + c$$

$$u = \frac{1}{4\mu} \frac{\partial p}{\partial X} r^2 - \frac{1}{4\mu} \frac{\partial p}{\partial X} R^2$$

$$u = \frac{1}{4\mu} \frac{\partial p}{\partial X} (r^2 - R^2)$$

$$u = -\frac{1}{4\mu} \frac{\partial p}{\partial X} (R^2 - r^2) \qquad \ldots \qquad \ldots \qquad (3)$$

Step 11

The ratio of maximum velocity to average velocity,

$$U_{max} = \bar{U}$$

$$\frac{U_{max}}{\bar{U}} = ?$$

Condition

If the velocity is maximum $U = U_{max}$, r = 0 sub in eq (3)

$$u = -\frac{1}{4\mu} \frac{\partial p}{\partial X} (R^2 - r^2)$$

$$u = -\frac{1}{4\mu} \frac{\partial p}{\partial X} (R^2)$$

$$\bar{U} = ?$$

Therefore discharge Q = A * $\bar{U}$

$$\bar{U} = \frac{Q}{A}$$

$$A = \pi R^2$$

Q = ?

Q = area elemental ring * velocity of the element

$$dQ = 2\pi r dr * U$$

$$dQ = 2\pi r dr * \left[-\frac{1}{4\mu} \frac{\partial p}{\partial X} (R^2 - r^2) \right]$$

Both Side Integrate

$$\int dQ = 2\pi \frac{\partial p}{\partial X} \left(-\frac{1}{4\mu} \right) \int_0^R (R^2 - r^2) r dr$$

$$Q = -\frac{\Pi}{2\mu} \frac{\partial p}{\partial X} \left[\frac{2R^4 - R^4}{4} \right]$$

$$Q = -\frac{\Pi}{2\mu} \frac{\partial p}{\partial X} \left[\frac{R^4}{4} \right]$$

$$Q = -\frac{\Pi}{8\mu} \frac{\partial p}{\partial X} \left[R^4 \right]$$

$$\bar{U} = \frac{Q}{\Pi R^2}$$

$$\bar{U} = \frac{-\frac{\Pi}{8\mu} \frac{\partial p}{\partial X} [R^4]}{\Pi R^2}$$

$$\bar{U} = -\frac{1}{8\mu} \frac{\partial p}{\partial X} [R^2]$$

$$\frac{U_{max}}{\bar{U}} = \frac{-\frac{1}{4\mu} \frac{\partial p}{\partial X} (R^2)}{-\frac{1}{8\mu} \frac{\partial p}{\partial X} [R^2]}$$

$$\frac{U_{max}}{\bar{U}} = 2$$

Drop of Pressure from the given Length of Pipe

$$\bar{U} = -\frac{1}{8\mu} \frac{\partial p}{\partial X} [R^2]$$

$$\frac{\partial p}{\partial X} = \bar{U} * -8\mu * \left[\frac{1}{R^2} \right]$$

$$\frac{\partial p}{\partial X} = -\frac{8\mu \bar{U}}{R^2}$$

$$\partial p = -\frac{8\mu \bar{U}}{R^2} \partial X$$

Integrate on both sides

$$\int_1^2 \partial p = -\frac{8\mu \bar{U}}{R^2} \int_1^2 \partial X$$

$$P_2 - P_1 = -\frac{8\mu \bar{U}}{R^2}$$

$$x_2 - x_1 = L$$

$$P_2 - P_1 = -\frac{8\mu \bar{U}}{R^2} L$$

$$h_f = \frac{P_1 - P_2}{\rho g}$$

$$h_f * \rho g = P_1 - P_2$$

$$h_f = \frac{P_1 - P_2}{\rho g} = \frac{8\mu\bar{U}}{R^2 \rho g}$$

$$h_f = \frac{8\mu\bar{U}}{\frac{D^2}{2}\,\rho g}$$

$$h_f = \frac{P_1 - P_2}{\rho g} = \frac{8\mu\bar{U}L * 4}{D^2 \rho g}$$

$$h_f = \frac{P_1 - P_2}{\rho g} = \frac{32\mu\bar{U}L}{D^2 \rho g}$$

2.3. Flow Through Pipe

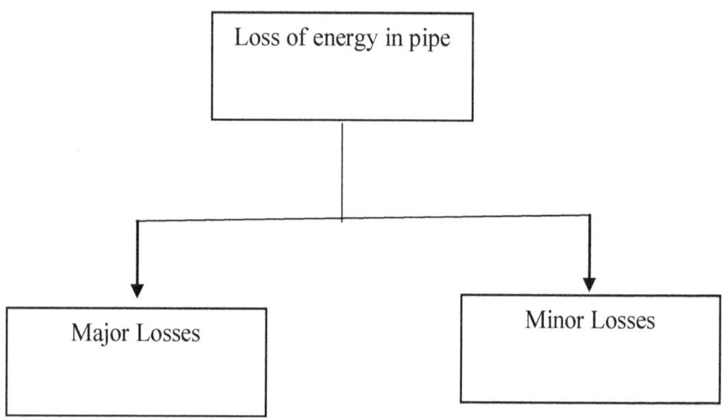

Major losses	Minor losses
This is due to friction. It is calculated by following formula	This is due to sudden expansion in pipe
1.Darcy weisbach formula	1. Sudden expansion in pipe
2. Chezy's equation	2. Sudden contraction in pipe
	3. Bending pipe
	4. An obstruction
	5. Pipe fitting

Table 2.1: Minor Losses

Loss of head due to sudden enlargement	Loss of head due to sudden contraction	Loss of head at entrance and exit	Loss of head due to obstruction in a pipe	Loss of head due to bend in pipe	Head loss in various pipe fittings
		Entrance Exit			--
h_{fc} $= \dfrac{(v_1 - v_2)^2}{2g}$ V_1 = velocity at the section 1-1 V_2 = velocity at the section 2-2	$h_{fc} = \dfrac{0.375v_2^2}{2g}$ [cc is given] $h_{fc} = \dfrac{0.5v_2^2}{2g}$ [cc is not given] Co-efficient of contraction	Entrance h_i $= \dfrac{0.5v_2^2}{2g}$ Exit $h_0 = \dfrac{v_2^2}{2g}$	$h_0 = \dfrac{v^2}{2g}\left[\dfrac{A}{c_c(A-a)} - 1\right]^2$ A – Area of the pipe (m²) A – area of obstruction(m²) $A = \dfrac{\Pi}{4}D^2$ D = diameter of the pipe $a = \dfrac{\Pi}{4}d^2$ d = diameter of the obstruction	h_b $= \dfrac{kv^2}{2g}$ K= co-efficient of bend	h_f $= \dfrac{kv^2}{2g}$ K= co-efficient of pipe fitting

2.4. Major Losses

1. Darcy weisbach formula

$$h_f = \frac{4flv^2}{2gD}$$

h_f = head loss due to friction (m)

l = length of the pipe (m)

v = velocity of flow

D= diameter of pipe

4f= co-efficient of friction

2. Chezy's equation

$$v = c\sqrt{mi}$$

Where

    C = chezy constant

    V = velocity of flow

    Q = a*v

    V = q/a

$$A = \frac{\Pi}{4}d^2$$

Where,

    d = diameter of the pipe

    m = hydraulic mean depth

    m = A/p = area/perimeter

    m = D/4

    i = loss of head per unit length of pipe

$$i = \frac{h_f}{L}$$

Darcy Formula

To find coefficient of friction 'f'

Case 1

$$f = \frac{16}{Re}\ [\text{Reynolds number Re} < 2000]$$

Case 2

$$f = \frac{0.0794}{\left(Re^{1/4}\right)}\ [\text{Reynolds number Re} > 4000]$$

$$\text{Reynolds number Re} = \frac{uD}{\gamma}$$

U = Velocity (m/s)

D = diameter (m)

γ = kinematic viscosity (m²/s)

2.5. Flow through Pipe in Series (Compound Pipe)

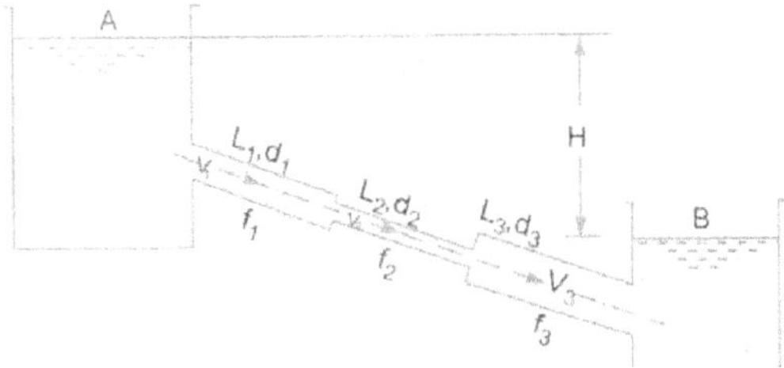

Figure 2.3: Compound Pipe

Step 1

It is defined as the pipe of differential line and different diameter are connected end to end (in series) to form a pipe as shown in figure

Step 2

The discharge passing through each pipe is same therefore

$$Q = A_1 V_1 = A_2 V_2 = A_3 V_3$$

Step 3

The Difference In Liquid surface level is equal to sum of total head loss in the pipe

$$H = \frac{0.5 V_1^2}{2g} + \frac{4 f_1 L_1 V_1^2}{2g d_1} + \frac{0.5 V_2^2}{2g} + \frac{4 f_2 L_2 V_2^2}{2g d_2} + \frac{(V_1 - V_2)^2}{2g} + \frac{4 f_3 L_3 V_3^2}{2g d_3} + \frac{V_3^2}{2g}$$

Step 4

If Minor Losses Are Neglected

$$H = \frac{4 f_1 L_1 V_1^2}{2g d_1} + \frac{4 f_2 L_2 V_2^2}{2g d_2} + \frac{4 f_3 L_3 V_3^2}{2g d_3}$$

Step 5

if the coefficient of friction is same

$$f_1 = f_2 = f_3 = f$$

$$H = \frac{4 f_1 L_1 V_1^2}{2g d_1} + \frac{4 f_2 L_2 V_2^2}{2g d_2} + \frac{4 f_3 L_3 V_3^2}{2g d_3}$$

Step 6

$$H = \frac{4f}{2g} \left[\frac{L_1 V_1^2}{d_1} + \frac{L_2 V_2^2}{d_2} + \frac{L_3 V_3^2}{d_3} \right]$$

2.6. Flow Through Pipe in Parallel

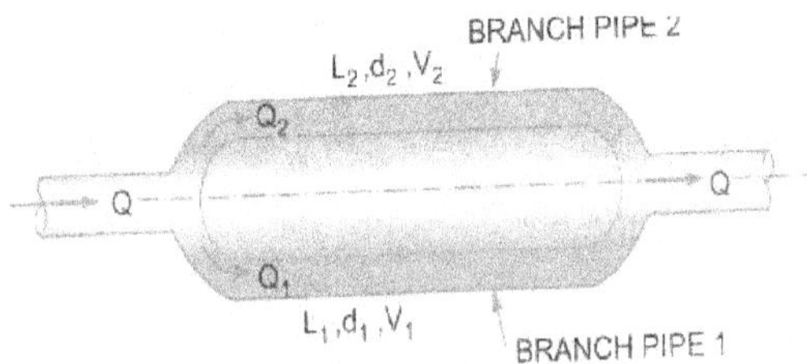

Figure 2.4: Flow through Pipe in Parallel

Step 1

Consider a main pipe which divides into two or more branches as shown in figure and again joined together to form a single pipe

Step 2

Therefore

Discharge $Q = Q_1 + Q_2$

Step 3

Loss Of Head In Pipe 1   =   Loss Head In Pipe 2

$$\frac{4fL_1V_1^2}{2gd_1} = \frac{4fL_2V_2^2}{2gd_2}$$

$$\frac{L_1V_1^2}{d_1} = \frac{L_2V_2^2}{d_2}$$

2.7. Equivalent Pipe and Dupuit's Equation

It is defined as the pipe of uniform diameter having loss of head and discharge = loss of head and discharge of a compound pipe consisting of several pipe of different length and diameter

The uniform diameter of the pipe is called equivalent pipe.

$$\frac{L}{d^5} = \frac{L_1}{d_1^5} + \frac{L_2}{d_2^5} + \frac{L_3}{d_3^5}$$

$$L = L_1 + L_2 + L_3$$

d_1 = diameter of pipe 1

d_2 = diameter of pipe 2

d_3 = diameter of pipe 3

Problems

1. Water flows at 0.05m³/s in a pipe of 20cm diameter, 500m long. Find the loss of head due to friction assuming friction factor is 0.025

Given Data

$Q = 0.05$m³/s

$d = 20$cm $= 0.2$m

$l = 500$m

$4f = 0.025$

$h_f =$

Solution

Step 1

Head loss due to friction $h_f = \dfrac{4fLv^2}{2gd}$

Step 2

$Q = A * V$

$V = Q/A = \dfrac{0.05 * 4}{\Pi*(0.2)^2} = 1.59$m/s

$h_f = \dfrac{0.025*500*(1.59)^2}{2*9.81*0.2} = \dfrac{31.69}{3.92}$

$h_f = 8.07m$

2. The rate of flow of water through horizontal pipe 0.25m³/s. the diameter of the pipe which is 200mm is suddenly enlarged to 400mm. the pressure intensity in the smaller pipe is 11.772 N/cm². Determine:

   a) Loss of head due to sudden enlargement

   b) Pressure intensity in a large pipe

   c) Power lost due to friction

Given Data

$Q = 0.25$m³/s (horizontal pipe)

$d_1 = 200$mm $= 0.2$m

$d_2 = 400$mm $= 0.4$m

$P_1 = 11.772$ N/cm² $= 11.772 *10^4$ N/m²

$P_2 = ?$

Solution

(i) Loss of head due to sudden enlargement

$$h_e = \frac{(v_1 - v_2)^2}{2g}$$

$$A_1 = \frac{\Pi}{4}d_1^2 = \frac{\Pi}{4}(0.2)^2 = 0.0314m^2$$

$$A_2 = \frac{\Pi}{4}d_2^2 = \frac{\Pi}{4}(0.4)^2 = 0.1256m^2$$

$$Q_1 = A_1 v_1$$

$$V_1 = \frac{Q_1}{A_1} = \frac{0.25}{0.0314} = 77.96 \ m/s$$

$$Q_2 = A_2 v_2$$

$$V_2 = \frac{Q_2}{A_2} = \frac{0.25}{0.1256} = 1.99 \ m/s$$

$$h_e = \frac{(v_1 - v_2)^2}{2g} = \frac{(7.96 - 1.99)^2}{2*9.81} = 1.816m$$

(ii) Pressure intensity in the larger pipe (P_2 =?)

Applying bernoullie theorem in section 1-1 and 2-2

$$\frac{P_1}{\rho g} + \frac{V_1^2}{2g} + Z_1 = \frac{P_2}{\rho g} + \frac{V_2^2}{2g} + Z_2 + h_e$$

$$\frac{(11.772 * 10^4)}{1000 * 9.81} + \frac{7.96^2}{2*9.81} = \frac{p_2}{1000 * 9.81} + \frac{1.99^2}{2*9.81} + 1.816$$

$$12 + 3.229 = \frac{p_2}{1000 * 9.81} + 0.202 + 1.816$$

$$p_2 = 12.96 * 10^4 \ \frac{N}{m^2}$$

(iii) Power lost

$$P = \frac{\rho g Q \, h_e}{1000} \ (m)$$

$$P = \frac{1000 * 9.81 * 0.25 * 1.816}{1000}$$

$$P = 4.48kw$$

3. A smooth pipe carries 0.30m³/s of water discharge with a head loss of 3m per 100m length of pipe if the water temperature is 200c. Determine the diameter of the pipe.

Given Data

Q = 0.30m³/s

h = 3m

L = 100m

T= 20⁰c

Solution

Head loss $h_f = \dfrac{4flv^2}{2gD}$

Q = A*V

$V = Q/A = \dfrac{0.3}{\left(\frac{\Pi}{4}\right)d^2} = \dfrac{1.2}{\Pi D^2}$

Assuming 4f =0.025

$$h_f = 0.025 * \dfrac{100}{2 * 9.81 * D}\left(\dfrac{1.2}{\Pi D^2}\right)^2$$

$$3 = \dfrac{3.6}{193.45D^5}$$

$$D^5 = 6.2 * 10^{-3}$$

$$D = 0.36m$$

4. Water is flowing through a horizontal pipe of diameter 200mm at velocity of 3m/s. the circular solid plate of diameter 150mm is placed in the pipe to obstruct the flow. Find the loss of head due to obstruction in the pipe if cc = 0.62

Given data

D = 200mm = 0.2m

V = 3m/s

d = 150mm = 0.15m

cc = 0.62

$h_{f\,ob}$ = ?

Solution

Head loss due to obstruction

$$h_f = \dfrac{v^2}{2g}\left[\dfrac{A}{cc(A-a)} - 1\right]^2$$

$$A = \frac{\Pi}{4}D^2 = \frac{\Pi}{4}(0.2)^2 = 0.0314m^2$$

$$a = \frac{\Pi}{4}d^2 = \frac{\Pi}{4}(0.15)^2 = 0.0177m^2$$

$$h_f = \frac{3^2}{2*9.81}\left[\frac{0.0314}{0.62*(0.0314-0.0177)} - 1\right]^2$$

$$h_f = 3.338m$$

5. At a sudden enlargement of a water main from 240mm to 480mm diameter the hydraulic gradient raises by 10mm estimate the rate of flow

Note

Hydraulic gradient is defined as the line which gives the sum of pressure head (P/W) and datum head (z) of a flowing fluid with respect to same reference line

$$HGL = \frac{P}{\rho g} + Z$$

Two sections 1-1 and 2-2

$$\frac{P_1}{\rho g} + Z_1 = \frac{P_2}{\rho g} + Z_2$$

$$\frac{P_1}{\rho g} + Z_1 - \frac{P_2}{\rho g} + Z_2 = HGL$$

Given data

d_1= 240mm = 0.24m

d_2=480mm = 0.48m

Hydraulic gradient rises = 10mm

Q=?

Solution

Applying Bernoulli's equation in section 1-1 and 2-2

$$\frac{P_1}{\rho g} + V + Z_1 = \frac{P_2}{\rho g} + Z_2$$

$$\left[\frac{P_2}{\rho g} + Z_2\right] - \left[\frac{P_1}{\rho g} + V + Z_1\right] = \frac{1}{100}m$$

$$h_e = (v_1 - v_2)^2$$

$$Q = A_1 V_1 = A_2 V_2$$

$$\frac{\Pi}{4}(0.24)^2 V_1 = \frac{\Pi}{4}(0.48)^2 V_2$$

$$V_1 = 4V_2$$

$$h_e = \frac{(4v_2 - V_2)^2}{2g} = \frac{9v_2^2}{2*9.81}$$

$$h_e = 0.452 V_2^2 \, (m)$$

$$\frac{v_1^2}{2g} - \frac{v_2^2}{2g} - h_e = \left[\frac{P_2}{\rho g} + z_2\right] - \left[\frac{P_1}{\rho g} + z_1\right]$$

$$\frac{(4v_2)^2}{2*9.81} - \frac{v_2^2}{2*9.81} - 0.452 v_2^2 = \frac{1}{100}$$

$$0.815 V_2^2 - 0.0509 V_2^2 - 0.452 V_2^2 = \frac{1}{100}$$

$$V_2^2 [0.3121] = \frac{1}{100}$$

$$V_2 = \frac{0.179 m}{s}$$

$$Q = A_2 V_2 = \frac{\Pi}{4}(0.48)^2 * 0.179$$

$$\frac{\Pi}{4}(0.0412)$$

$$Q = 0.032 m^3/s$$

6. A horizontal pipe of diameter 500mm is suddenly contracted to a diameter of 250mm the pressure intensity is in the large and smaller pipe is given as 13.73N/cm² respectively. Find the loss of head due to contraction if CC=0.62 also determine the rate of flow.

Given data

Horizontal pipe $Z_1 = Z_2$

d_1=500mm = 0.5m

d_2=250mm=0.25m

P_1=13.773 N/cm²

P_2=11.772 N/cm²

CC = 0.62

Q =?

H_{fc} =?

Solution

Applying Bernoulli's equation

$$\frac{P_1}{\rho g} + \frac{V_1^2}{2g} + Z_1 = \frac{P_2}{\rho g} + \frac{V_2^2}{2g} + Z_2 + h_f$$

$$\frac{P_1}{\rho g} + \frac{V_1^2}{2g} + Z_1 = \frac{P_2}{\rho g} + \frac{V_2^2}{2g} + Z_2 + h_c$$

$$h_c = \frac{0.375\,V_2^2}{2g}$$

Continuity equation

$$Q = A_1 V_1 = A_2 V_2$$

$$\frac{\Pi}{4}(0.5)^2\ V_1 = \frac{\Pi}{4}(0.25)^2 V_2 \quad V_1 = 0.25 V_2$$

$$V_2 = 4V_1$$

$$h_c = 0.375\frac{4V_1^2}{2g}$$

$$h_c = 0.306 V_1^2\,(m)$$

$$\frac{13.73*10^4}{1000*9.81} + \frac{V_1^2}{2*9.81} = \frac{11.772*10^4}{1000*9.81} + \frac{16v_1^2}{2*9.81} + 0.306(v_1^2)$$

$$V_1^2[0.0509 - 0.815 - 0.306] = 12 - 14 \ , \ V_1^2[-1.07] = -2$$

$$v_1^2 = \frac{2}{1.07}$$

$$V_1 = 1.367\,\frac{m}{s}$$

$$h_c = 0.306 V_1^2$$

$$h_c = 0.572 m$$

$$Q = A_1 V_1 = \frac{\Pi}{4}(0.5)^2(1.367) = \frac{\Pi}{4}(0.3417)$$

$$Q = 0.268\,\frac{m^3}{s}$$

7. A 150mm diameter pipe reduce in diameter abruptly to 100mm diameter if the pipe carries water at 30 l/sec calculate the pressure loss across contraction take CC =0.6

Given data

d_1 = 150mm = 0.15m

d_2 = 100mm = 0.1m

Q = 30 lit/sec = 30/1000 m³/s

Cc=0.6

Pressure lost =?

Solution

$$area\ A_1 = \frac{Q}{V_1} == V_1 = \frac{Q}{A_1} = \frac{0.03}{\left(\frac{\Pi}{4}\right) * (0.15)^2}$$

$$V_1 = 1.6989\,\frac{m}{s}$$

$$area\ A_2 = \frac{Q}{V_2} == V_2 = \frac{Q}{A_2} = \frac{0.03}{\left(\frac{\Pi}{4}\right) * (0.1)^2}$$

$$V_2 = 3.82\,\frac{m}{s}$$

Bernoulli's Equation

$$\frac{P_1}{\rho g} + \frac{V_1^2}{2g} + Z_1 = \frac{P_2}{\rho g} + \frac{V_2^2}{2g} + Z_2 + h_c$$

$$h_c = 0.375\frac{V_2^2}{2g} = \frac{0.375(3.82)^2}{2 * 9.81} = 0.278\ m$$

$$\frac{P_1}{\rho g} + \frac{V_1^2}{2g} + Z_1 = \frac{P_2}{\rho g} + \frac{V_2^2}{2g} + Z_2 + h_c$$

$$P_1 - P_2 = \rho g\left[\frac{V_2^2}{2g} - \frac{V_1^2}{2g} + h_c\right.$$

$$= 1000 * 9.81\left[\frac{3.82^2}{2*9.81} - \frac{1.698^2}{2*9.81} + 0.278\right]$$

$$P_1 - P_2 = 8583.75$$

8. Three pipes of length 800m, 500m, and 400m and of diameter 500mm, 400mm and 300mm respectively are connected in series. These pipes are to be replaced by a single pipe of length 1700m. Find the diameter of single pipe.

Given data

Identification

L_1, L_2, L_3 and $L = 4$ lengths

$d_1, d_2, d_3 = 3$ diameters

$d =?$

$L_1 = 800m \qquad d_1 = 0.5m$

$$L_2 = 500m \qquad d_2 = 0.4m$$
$$L_3 = 400m \qquad d_3 = 0.3m$$

Solution

Dupuit's equation is

$$\frac{L}{d^5} = \frac{L_1}{d_1^5} + \frac{L_2}{d_2^5} + \frac{L_3}{d_3^5}$$

$$\frac{1}{d_5} = \frac{1}{1700}\left[\frac{800}{0.5^5} + \frac{500}{0.4^5} + \frac{400}{0.3^5}\right]$$

$$\frac{1}{d^5} = 140.61$$

$$d_5 = 0.071$$

$$d = 0.3718m$$

Problems on Parallel Pipes

1. A main pipe divided into two parallel pipe which again forms one pipe length and diameter for the first parallel pipe are 200mm and 1m respectively, while the length and diameter of second parallel pipe are 2000m and 0.8m. Find the rate of flow in each parallel pipe if total flow in each parallel pipe if total flow in the main is 3m/s the co-efficient of friction for each parallel pipe is same and equal 0.005.

Given data

Parallel pipe

L_1 = 2000m

d_1 = 1m

L_2 = 2000m

d_2 = 0.8m

Q = 3 m³/s

4f =0.005

Solution

Discharge

$$Q = Q_1 + Q_2$$

$$Q_1 = A_1 V_1 = \frac{\Pi}{4} d_1^2 V_1 = 0.785 V_1 \frac{m^3}{s}$$

$$Q_2 = A_2 V_2 = \frac{\Pi}{4} d_2^2 V_2 = 0.503 V_2 \frac{m^3}{s}$$

Loss of head in pipe 1 = loss of head in pipe 2

$$\frac{4fL_1 V_1^2}{2gd_1} = \frac{4fL_2 V_2^2}{2gd_2}$$

$$\frac{L_1 V_1^2}{d_1} = \frac{L_2 V_2^2}{d_2}$$

$$\frac{2000 V_1^2}{1} = \frac{2000 V_2^2}{0.8}$$

$$V_1^2 = 1.25 V_2^2$$

$$V_1 = 1.118 \, V_2 \frac{m}{s}$$

$$Q = Q_1 + Q_2$$
$$3 = 0.7785(V_1) + 0.503(V_2)$$
$$3 = [0.78776 + 0.503]V_2 \quad (\text{sub } V_1 = 1.118 V_2)$$

$$3 = 1.38 V_2$$

$$V_2 = 2.17 \frac{m}{s}, \, V_1 = 2.429 \frac{m}{s}$$

$$Q_1 = A_1 V_1 = 0.785(2.429) = 1.90 \frac{m^3}{s}$$

$$Q_2 = A_2 V_2 = 0.503(2.17) = 1.09 \frac{m^3}{s}$$

2. A pipe of diameter 20cm and length 2000m connects two reservoir, having difference of water level has 20m. Determine the discharge through the pipe. If an additional pipe of diameter 20cm and length 1200m is attached to the last 1200m length of the existing pipe. Find the increase in the discharge. Take f=0.015 and neglect minor losses

Solution

Case 1

$$h_f = H = \frac{4flV^2}{2gd}$$

$$20 = \frac{4*0.015*2000*V^2}{2*9.81*0.2}$$

$$V = 0.808 \frac{m}{s}$$

$$Q = A * V = \frac{\Pi}{4}d^2 * V = \frac{\Pi}{4}(0.2)^2 * 0.808 = 0.025 \frac{m^3}{s}$$

Case II

Discharge $Q_1 = Q_2 + Q_3$ [diameter are same Q₂=Q₃]

$$Q_1 = 2Q_2$$

$$H = \frac{4f L_1 V_1^2}{(2gd_1)} + \frac{4f L_2 V_2^2}{(2gd_2)}$$

$$20 = \frac{4*0.015*800*V_1^2}{2*9.81*0.2} + \frac{4*0.015*1200*V_2^2}{2*9.81*0.2}$$

$$20 = 12.23V_1^2 + 18.35V_2^2$$

$$V_1 = \frac{Q_1}{A_1} = \frac{Q_1}{\frac{\Pi}{4}*(0.2)^2} = 31.847 Q_1 \frac{m}{s} \qquad\qquad 1$$

$$V_2 = \frac{Q_2}{A_2} = \frac{Q_2}{\frac{\Pi}{4}*(0.2)^2} = 15.92 Q_1 \frac{m}{s}$$

$$20 = 12.23(31.847 Q_1^2) + 18.35(15.92 Q_1^2)$$

$$Q_1 = 0.034 \frac{m^3}{s}$$

$$increase\ in\ discharge = Q_1 - Q = 0.009 \frac{m^3}{s}$$

Two Marks Question and Answers

1. Write down Hagen Poiseuille equation for Laminar fluid.

$$P_1 - P_2 = \frac{32\mu\, v\, L}{D^2}$$

   μ - Viscosity; v – Velocity; L – Length; D – Diameter.

2. Mention any three applications of Bernoulli's theorem.

   a) Venturimeter.

   b) Orifice meter.

   c) Pitot tube

3. **Compare Steady and Unsteady flow**

Steady flow	Unsteady flow
Steady flow is defined as that type of flow in which the fluid characteristics like Velocity, Pressure, Density etc., at a point do not change with time.	Unsteady flow is that type of flow in which the velocity, pressure, density etc., at a point changes with respect to time.
$\dfrac{\partial V}{\partial t} = 0; \dfrac{\partial p}{\partial t} = 0; \dfrac{\partial \rho}{\partial t} = 0$	$\dfrac{\partial V}{\partial t} \neq 0; \dfrac{\partial p}{\partial t} \neq 0; \dfrac{\partial \rho}{\partial t} \neq 0$

4. **Compare Uniform flow and Non uniform flow**

Uniform flow	Non-uniform flow
Uniform flow is defined as that type of flow in which the velocity at any given time does not change with respect to space.	Non Uniform flow is defined as that type of flow in which the velocity at any given time does change with space.
$\dfrac{\partial V}{\partial S} = 0;$	$\dfrac{\partial V}{\partial S} \neq 0;$

5. **Compare Laminar flow and Turbulent flow**

Laminar flow	Turbulent flow
Laminar flow is defined as that type of flow in which fluid particles move along well-defined path. Re < 2000 is called laminar flow.	Turbulent flow is defined as that type of flow in which the particle move in a Zig Zag way.

6. **Compare Compressible flow and Incompressible flow.**

Compressible flow	Incompressible flow
Compressible flow is that type of flow in which the density of the fluid changes from point to point or in other words the density is not constant.	Incompressible flow is that type of flow in which the density is constant for the fluid.
$\rho \neq$ Constant	$\rho =$ Constant.

7. Define Discharge and give the units.

It is defined as Quantity of water delivered per seconds.

Q = Area x Velocity = A x v (m³/s)

8. Give the continuity equation for one dimensional flow

$Q = A_1V_1 = A_2V_2$

9. Name the different forces present in fluid flow.

   a) Inertia force.

   b) Viscous force.

   c) Surface tension force.

   d) Gravity force.

10. Compare Stream line and Path line

Stream line	Path Line
A stream line is an imaginary line drawn through a flowing fluid in such a way that the tangent at once point on it indicates the velocity at that point.	A path line that is traced by a single fluid particle as it moves over a period of time.

11. State Bernoulli's theorem

Bernoulli's theorem states an ideal incompressible fluid when the flow is steady and continuous the sum of pressure, pressure energy, Kinematic energy and potential energy is constant along the stream line.

$$\frac{P}{\rho g} + \frac{v^2}{2g} + Z = \text{Constant}$$

$\dfrac{P}{\rho g}$ = Pressure energy; $\dfrac{v^2}{2g}$ = Kinematic energy; Z = Potential energy

12. What are the assumptions made on Bernoulli's equation?

   1. The fluid is ideal.

   2. The flow is steady.

   3. The flow is incompressible.

   4. The flow is irrotational.

Incompressible Fluid Flow

13. Write down Hagon-Poiseuille equation for Laminar flow.

$$P_1 - P_2 = \frac{32\mu UL}{D^2} = \frac{128\mu QL}{\pi D^4}$$

P_1, P_2 = Pressure Inlet and Outlet; μ = Dynamic Viscosity

V = Velocity, L = Length, D = Diameter; Q = Discharge

14. Difference between Total energy Line and hydraulic gradient lines.

Total energy line If at different sections of the pipes total energy is plotted to scale and joined by a line, the line is called energy grade line.	Hydraulic gradient line (HGL). The pressure head in a pipe decreased gradually from section to section of pipe in the direction of fluid flow due to loss of energy. If pressure heads at different sections of the pipe are joined by a straight line. This line is called hydraulic grade line or pressure line.
T.E. = Pressure head of datum head / Kinetic head.	H.G.L. = Pressure head + datum head.

15. What is a siphon? What are its applications?

A siphon is a long bend pipe used for carrying water from a reservoir at a higher head to another reservoir at a lower head when the two reservoirs are by separated by a hill.

16. What are major and minor losses of flow through pipes?

Major Losses	Minor Losses

This is due to friction and is calculated by the following formula.
 	
  a. Darcy Weisbach formula
  b. Chercy formula
	1. Sudden expansion of pipe.
	2. Sudden contraction of pipe
	3. Bend in pipe
	4. Pipe fittings
	5. An obstruction in pipe

17. What is Reynold's number? And what is the physical significance of Reynold's number?

$$\text{Reynolds numbers (Re)} = \frac{Inertia\ force}{Viscous\ force} \quad OR \quad \frac{U\ D}{\gamma}$$

18. Mention the significance of Reynolds model law.

    1. Motion of air planes.

    2. Motion of submarines.

    3. Pipe flow.

    4. Flow of incompressible fluid in closed pipes.

19. What is similarities in model study?

    1. Geometric similarity.

    2. Kinematic similarity.

    3. Dynamic similarity.

20. Define Drag and Lift

Drag	Lift
The component of total force in the direction of flow of fluid is known as Drag.	The component of total force in the direction perpendicular to the direction of flow is known as Lift.

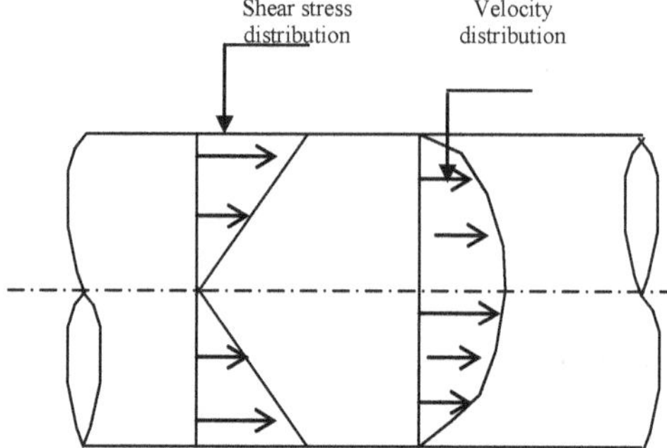

21. Sketch velocity distribution curves for Laminar flow in pipes and sketch the shear stress distribution curves.

22. Write down two examples of Laminar flow.

    1.   Flow of Viscous fluid through circular pipe.

    2.   Flow of viscous fluid between two parallel plates.

23. What are pipes in series?

It is defined as the pipes of different chambers and length are connected with one another to form a single pipe line.

24. What do you meant by flow through parallel pipes?

When a main pipe line divides into two or more parallel pipes, which again join together to form a single pipe and continue as a main line. These pipes are said to be pipes in parallel.

25. What is equivalent pipe?

A compound pipe consisting of several pipe of varying diameters and length may be replaced by a pipe of uniform diameter, which is known as equivalent pipe.

26. Write the formula for Darcy Weisbach equation.

$$h_f = \frac{4fLV^2}{2gD}$$

f = Coefficient of friction; L = Length; V = Velocity of fluid; D = Diameter of pipe.

27. Define Displacement thickness.

It is defined as the distance-measured perpendicular to the boundary by which main stream is displaced to on account of formation of boundary layer.

Unit II

Flow through Circular Conduits

Part-A

1. What is the physical significance of Reynold's number?
2. Define boundary layer and give its significance.
3. List the causes of minor energy losses in flow through pipes.
4. What is T.E.L.?
5. What is Hydraulic Gradient Line?
6. Write down Hagen-Poiseuille equation for laminar flow
7. What are energy lines and hydraulic gradient lines?
8. Write down four examples of laminar flow.
9. What between laminar and turbulent flow.
10. What is a syphon? What are its applications?
11. What are the losses experienced by a fluid when it is passing through a pipe?
12. What is equivalent pipe?
13. What do you mean by flow through parallel pipes
14. Mention the range of Reynold's number for laminar and turbulent flow in a pipe.
15. What is the difference between a laminar flow and turbulent flow?
16. In laminar flow through a pipe the maximum velocity at the pipe axis is 0.2m/s. Find the average velocity.
17. Define boundary layer thickness.
18. Define displacement thickness.
19. Define momentum thickness.
20. What are hydraulic gradient lines?

Part - B

1. The velocity distribution in the boundary layer is given by $u/U = y/\delta$, where u is the velocity at a distance y from the plate u=U at y =δ, δ being boundary layer thickness. Find the displacement thickness, momentum thickness and energy thickness.
2. Derive an expression for head loss through pipes due to friction.
3. Explain the losses of energy in flow through pipes.
4. Determine the equivalent pipe corresponding to 3 pipes in series with lengths and diameters L1, L2, L3, d1, d2, d3 respectively.

5. The rate of flow of water through a horizontal pipe is 0.25 ms³/sec. The diameter of the pipe which is 20 cm is suddenly enlarged to 40 cm. The pressure intensity in the smaller pipe is 11.772 N/cm².

Determine:

a. Loss of head due to sudden enlargement.

b. Pressure intensity in larger pipe.

c. Power loss due to enlargement.

6. An oil of sp.gravity 0.7 is flowing through a pipe of diameter 30 cm at the rate of 500 litres/sec. Find the head lost due to friction and power required to maintain the flow for a length of 1000 m. Take υ= 0.29 stokes.

7. For a town water supply, a main pipe line of diameter 0.4 m is required. As pipes more than 0.35m diameter are not readily available, two parallel pipes of same diameter are used for water supply. If the total discharge in the parallel pipe is same as in the single main pipe, find the diameter of parallel pipe. Coefficient of discharge to be the same for all the pipes.

8. A pipe line 10km, long delivers a power of 50 kW at its outlet ends. The pressure at inlet is 500 kN/m² and pressure drop per km of pipeline is 50 kN/m. Find the size of the pipe and efficiency of transmission. Take 4f= 0.02.

9. The velocity of water in a pipe 200mm diameter is 5m/s. The length of the pipe is 50m. Find the loss of head due to friction, if f= 0.08.

10. A power transmission pipe 10 cm diameter and 500 m long is fitted with a nozzle at the exit, the inlet is from a river with water level 60 m above the discharge nozzle. Assume f = 0.02, calculate the maximum power which can be transmitted and the diameter of nozzle required

UNIT III

DIMENSIONAL ANALYSIS

3.1.   Fundamental Dimensions: (MLT)

Symbol   Unit

Mass (M)   Kg

Length (L)   m

Time (T)   s

3.2.   Dimensional Homogenity

Both sides of the dimension in the equation are same

LHS   =   RHS

$$V \quad = \quad Cv \times \sqrt{2gH}$$

By applying the dimension and unit.

$$\frac{m}{s} \quad = \quad \sqrt{\frac{m}{s^2}} \ x \ m \ = \sqrt{\frac{m^2}{s^2}}$$

$$\frac{m}{s} \quad = \quad \frac{m}{s}$$

$$\frac{L}{T} \quad = \quad \frac{L}{T}$$

$$LT^{-1} \quad = \quad LT^{-1}$$

Hence, it is proved.

3.3.   Methods of Dimensional Analysis

If the number of variable involved in a physical phenomenon are known, then the relation among the variables can be determined by the following two methods:

1. Rayleigh's method, and
2. Buckingham's π-theorem

3.4. Rayleigh's Method

- This method is used for determining the expression for a variable which depends upon maximum three or four variables only. If the number of independent variables becomes more than four, then it is very difficult to find the expression for the dependent variable.

- Let, X is a variable, which depends on X_1, X_2 and X_3 variables. Then according to Rayleigh's method, X is function of X_1, X_2 and X_3 and mathematically it is written as X = f $[X_1, X_2, X_3]$.

- This can also be written as $X = KX_1^a, X_2^b, X_3^c$

- The values of a, b and c are obtained by comparing the powers of the fundamental dimension on both sides. Thus the expression is obtained for dependent variable.

3.5. Buckingham's π -Theorem

If they are 'n' variables physical phenomenon and if this variables contains 'm' fundamental dimension (MLT). Then the variables are arranged in to 'n-m' dimension less than each term is called as π term.

Step 1

Let X_1, X_2, X_3 ...... X_n are the variable involved in a physical problem. Let X_1 be the dependent variable and X_2, X_3 .... X_n are the independent variables on which X_1 depend. Then X_1 is a function of X_2, X_3 .... X_n and mathematically it is expressed as

$$X_1 = f(X_2, X_3, ... X_n) \qquad\qquad .... \qquad (i)$$

Equation (i) can also be written as

$$f_1 (X_1, X_2, X_3, ... X_n) \quad = \quad 0 \qquad\qquad .... \qquad (ii)$$

Step 2

Equation (ii) is a dimensionally homogeneous equation. It contains n variables. If there are m fundamental dimensions then according to Buckingham's π-theorem, equation (ii) can be written in terms of number of dimensionless groups or π-terms in which number of π-terms is equal to (n - m). Hence, equation (ii) becomes as

$$f(\pi_1, \pi_2 ..... \pi_{n-m}) \quad = \quad 0 \qquad\qquad (iii)$$

Step 3

Each of π-terms is dimensionless and is independent of the system. Division or multiplication by a constant does not change the character of the π-term. Each π-term contains m + 1 variables, where m is the number of fundamental dimensions and is also called repeating variables. Let in the above case X_2, X_3 and X_4 are repeating variables if the fundamental dimension m (M, L, T) = 3. Then each π – term is written as

$$
\left.
\begin{aligned}
\pi_1 &= X_2{}^{a_1}, X_3{}^{b_1}.X_4{}^{c_1}.X_1 \\
\pi_2 &= X_2{}^{a_2}.X_3{}^{b_2}.X_4{}^{c_2}.X_5 \\
&\quad \cdot \\
&\quad \cdot \\
&\quad \cdot \\
\pi_{n-m} &= X_2{}^{a_{n-m}}.X_3{}^{b_{n-m}}.X_4{}^{c_m}.X_n
\end{aligned}
\right\} \qquad \cdots \qquad \text{(iv)}
$$

Step 4

Each equation is solved by the principle of dimensional homogeneity and values of a_1, b_1, c_1 etc., are obtained. These values are substituted in equation (iv) and values of π_1, π_2, …. π_{n-m} are obtained. These values of π' are substituted in equation (iii). The final equation for the phenomenon is obtained by expressing any one of the π-terms as a function of others as

$$
\begin{aligned}
\pi_1 &= \phi\left[\pi_2, \pi_3, \ldots \pi_{n-m}\right] \\
\pi_2 &= \phi_1\left[\pi_1, \pi_3, \ldots \pi_{n-m}\right]
\end{aligned}
\qquad \cdots \qquad \text{(v)}
$$

3.6. Methods of Selecting the Repeating Variables

The number of repeating variables are equal to the number of fundamental dimensions of the problem. The choice of repeating variables if governed by the following considerations:

1. As far as possible, the dependent variable should not be selected as repeating variable.
2. The repeating variables should be chosen in such a way that one variable contains geometric property, other variable contains flow property and third variable contains fluid property.

Geometric property	Flow properties	Fluid properties
Length (l) Diameter (d) Height (h)	velocity (v) angular velocity (ω) acceleration (a) angular acceleration (α) discharge (q) speed (N)	Density (P) Dynamic Viscosity (μ) Kinematic Viscosity (Γ)
L D H	V ω a α Q N	ρ μ γ

1. State Buckingham's π-theorem and explain step by step procedures for Buckingham's π-theorem with example.

Solution

Statement

If they are 'n' variables physical phenomenon and if this variables contains 'm' fundamental dimension (MLT). Then the variables are arranged in to 'n-m' dimension less than each term is called as π term.

Example

A Frictional torque (τ) of the diameter 'D' and rotated at a speed 'N' in a fluid of viscosity (μ) and density (ρ) in a turbulent flow is given by $T = D^5 N^2 \rho \, \phi \left[\dfrac{\mu}{D^2 N \rho} \right]$ using Buckingham's π-theorem

Step by Step Procedure

Step 1

Total number of variables n=5

Step 2

Fundamental dimension m = 3 (MLT)

Step 3

n - m = 5 – 3 = 2 (therefore 2 π-term)

Step 4

	Repeating variables	**Non-repeating variables**	
π_1	$D^{a_1} N^{b_1} \rho^{c_1}$	T	→ 1
π_2	$D^{a_2} N^{b_2} \rho^{c_2}$	μ	→ 2

Step 5

Applying dimensions

$$\pi_1 = M^0 L^0 T^0 = (L)^{a_1} (T^{-1})^{b_1} (ML^{-3})^{c_1} \, ML^2 T^{-2}$$

$$\pi_2 = M^0 L^0 T^0 = (L)^{a_2} (T^{-1})^{b_2} (ML^{-3})^{c_2} \, ML^{-1} T^{-1}$$

Step 6

$$\pi_1 = M^0\,L^0\,T^0 = (L)^{a_1}\,(T^{-1})^{b_1}\,(ML^{-3})^{c_1}\,ML^2\,T^{-2}$$

Equating power of 'm'	Equating power of 'L'	Equating power of 'T'
$0 = c_1 + 1$ $c_1 = -1$	$0 = a_1 - 3c_1 + 2$ $0 = a_1 - 3(1) + 2$ $a_1 = -5$	$0 = -b_1 - 2$ $b_1 = -2$

a_1, b_1, c_1 sub in eq 1

$$\pi_1 \;=\; D^{-5}\,N^{-2}\,\rho^{-1}\,T$$

$$\pi_1 \;=\; \frac{T}{D^5 N^2 \rho}$$

Step 7

$$\pi_2 = M^0\,L^0\,T^0 = (L)^{a_2}\,(T^{-1})^{b_2}\,(ML^{-3})^{c_2}\,ML^{-1}\,T^{-1}$$

Equating power of 'm'	Equating power of 'L'	Equating power of 'T'
$0 = c_2 + 1$ $c_2 = -1$	$0 = a_2 - 3c_2 - 1$ $0 = a_2 - 3(-1) - 1$ $a_2 = -2$	$0 = -b_2 - 1$ $b_2 = -1$

a_2, b_2, c_2 sub in eqn. 2

$$\pi_2 \;=\; D^{-3}\,N^{-1}\,\rho^{-1}\,\mu$$

$$\pi_2 \;=\; \frac{\mu}{D^2 N^1 \rho}$$

Step 8

$$\pi_1 = \varphi\,[\pi_2]$$

$$\frac{T}{D^5 N^2 \rho^1} = \phi\left[\frac{\mu}{D^2 N^2 \rho^1}\right]$$

$$T = D^5 N^2 \rho \;\; \phi\left[\frac{\mu}{D^2 N^2 \rho}\right] \qquad \text{Hence Proved}$$

1. Using Buckingham's π-Theorem show that the discharge q consumed by an oil ring is given by

$$Q = Nd^3\,\phi\left\{\frac{\mu}{\rho Nd^2},\,\frac{\sigma}{\rho N^2 d^3},\,\frac{\omega}{\rho N^2 d}\right\} \text{ when d is the diameter of the ring, N is speed, } \rho \text{ is Density, } \mu \text{ is}$$

Viscosity, σ is surface tension, ω is specific weight.

Solution

Step 1

Total number of variables n = 7

Step 2

Fundamental dimension m = 3 (MLT)

Step 3

n - m = 7 – 3 = 4 (therefore 4 π-term)

Step 4

EQ		Repeating variables	Non-repeating variables
1	π_1	$\rho^{a_1} N^{b_1} d^{c_1}$	Q
2	π_2	$\rho^{a_2} N^{b_2} d^{c_2}$	μ
3	π_3	$\rho^{a_3} N^{b_3} d^{c_3}$	σ
4	π_4	$\rho^{a_4} N^{b_4} d^{c_4}$	ω

Step 5

Applying dimensions

$$\pi_1 = m^0\, L^0\, T^0 = \qquad \rho^0\, N\, d \quad Q \qquad = (1)^{a_1}\, (T^{-1})^{b_1}\, (ML^{-3})^{c_1} \quad L^3 T^{-1}$$

$$\pi_2 = m^0\, L^0\, T^0 = \qquad \rho\, N\, d \quad μ \qquad = (ML^{-3})^{a_2}\, (T^{-1})^{b_2}\, (L)^{c_2} \quad ML^{-1} T^{-1}$$

$$\pi_3 = m^0\, L^0\, T^0 = \qquad \rho\, N\, d \quad σ \qquad = (ML^{-3})^{a_3}\, (T^{-1})^{b_3}\, (L)^{c_3} \quad MT^{-2}$$

$$\pi_4 = m^0\, L^0\, T^0 = \qquad \rho\, N\, d \quad ω \qquad = (ML^{-3})^{a_4}\, (T^{-1})^{b_4}\, (L)^{c_4} \quad ML^{-2} T^{-2}$$

Step 6

$$\pi_1 = m^0\, L^0\, T^0 = \rho^0\, N\, d\, Q = (1)^{a_1}\, (T^{-1})^{b_1}\, (ML^{-3})^{c_1} \quad L^3 T^{-1}$$

Equating power of 'm'	Equating power of 'L'	Equating power of 'T'
$0 = a_1$	$0 = c_1 + 3$	$0 = -b_1 + (-1)$
$a_1 = 0$	$C_1 = -3$	$b_1 = -1$

a_1, b_1, c_1 sub in eq 1

$$\pi_1 = \rho^{a_1} N^{b_1} d^{c_1} Q$$

$$\pi_1 = \frac{Q}{Nd^3}$$

Step 7

$$\pi_2 = m^0\, L^0\, T^0 = \rho\, N\, d μ = (ML^{-3})^{a_2}\, (T^{-1})^{b_2}\, (L)^{c_2} \quad ML^{-1} T^{-1}$$

Equating power of 'm'	Equating power of 'L'	Equating power of 'T'
	$0 = -3a_2 + c_2 - 1$	
$0 = a_2 + 1$	$0 = -3(-1) + c_2 - 1$	$0 = -b_2 - 1$
$a_2 = -1$	$0 = 3 + c_2 - 1$	$b_2 = -1$
	$c_2 = -2$	

a_2, b_2, c_2 sub in eqn. 2

$\pi_2 = \rho^{\,a_2}\, N^{\,b_2}\, d^{\,c_2}\, \mu$

$$\pi_2 = \frac{\mu}{\rho N d^2}$$

Step 8

$\pi_3 = m^0\, L^0\, T^0 = \rho\, N\, d\, \sigma = (ML^{-3})^{a_3}\, (T^{-1})^{b_3}\, (L)^{c_3}\, MT^{-2}$

Equating power of 'm'	Equating power of 'L'	Equating power of 'T'
$0 = a_3 + 1$ $a_3 = -1$	$0 = -3a_3 + c_3$ $0 = -3(-1) + c_3$ $0 = 3 + c_3$ $C_3 = -3$	$0 = -b_3 - 2$ $b_3 = -2$

a_3, b_3, c_3 sub in eq 2

$\pi_3 = \rho^{\,a_3}\, N^{\,b_3}\, d^{\,c_3}\, \sigma$

$$\pi_3 = \frac{\sigma}{\rho N^2 d^3}$$

<Step 9>

$\pi_4 = m^0\, L^0\, T^0 = \rho\, N\, d\, \omega = (ML^{-3})^{a_4}\, (T^{-1})^{b_4}\, (L)^{c_4}\, ML^{-2}T^{-2}$

Equating power of 'm'	Equating power of 'L'	Equating power of 'T'
$0 = a_4 + 1$ $a_4 = -1$	$0 = -3a_4 + c_4 - 2$ $0 = -3(-1) + c_4 - 2$ $0 = 3 - 2 + c_4$ $C_4 = -1$	$0 = -b_4 - 2$ $b_4 = -2$

a_4, b_4, c_4 sub in eqn. 4

$\pi_4 = \rho^{\,a_4}\, N^{\,b_4}\, d^{\,c_4}\, \omega$

$$\pi_4 = \frac{\omega}{\rho N^2 d}$$

<Step 10>

$\pi_1 = \varphi\,[\pi_2,\, \pi_3,\, \pi_4,]$

$$\frac{Q}{Nd^3} = \varphi\left\{ \frac{\mu}{\rho Nd^2},\, \frac{\sigma}{\rho N^2 d^3},\, \frac{\omega}{\rho N^2 d} \right\}$$

$$Q = Nd^3 \varphi\left\{ \frac{\mu}{\rho Nd^2},\, \frac{\sigma}{\rho N^2 d^3},\, \frac{\omega}{\rho N^2 d} \right\} \qquad \text{Hence Proved}$$

2. Using Buckingham π theorem show that velocity through circular orifices is given by $V = \sqrt{2gH}\, f\left(\dfrac{D}{H},\, \dfrac{\mu}{\rho VH} \right)$ where V = velocity through orifice, D = diameter, H = head casing

flow and ρ, μ are density and dynamic viscosity of the passing through orifice and g is acceleration due to gravity.

Solution

Step 1

Total number of variables n = 6

Step 2

Fundamental dimension m = 3 (MLT)

Step 3

n - m = 6 – 3 = 3 (therefore 3 π-term)

Step 4

EQ		Repeating variables	Non-repeating variables
1	π_1	$H^{a_1} g^{b_1} \rho^{c_1}$	V
2	π_2	$H^{a_2} g^{b_2} \rho^{c_2}$	D
3	π_3	$H^{a_3} g^{b_3} \rho^{c_3}$	μ

Step 5

Applying dimensions

$\pi_1 = m^0 L^0 T^0 = \quad Hg\rho \quad V \quad = (L)^{a_1} (LT^{-2})^{b_1} (ML^{-3})^{c_1} \ LT^{-1}$

$\pi_2 = m^0 L^0 T^0 = \quad Hg\rho \quad D \quad = (L)^{a_2} (LT^{-2})^{b_2} (ML^{-3})^{c_2} \ L$

$\pi_3 = m^0 L^0 T^0 = \quad Hg\rho \quad \mu \quad = (L)^{a_3} (LT^{-2})^{b_3} (ML^{-3})^{c_3} \ ML^{-1}T^{-1}$

Step 6

$\pi_1 = m^0 L^0 T^0 = Hg\rho V = (L)^{a_1} (LT^{-2})^{b_1} (ML^{-3})^{c_1} \ LT^{-1}$

Equating power of 'm'	Equating power of 'L'	Equating power of 'T'
$0 = C_1$ $C_1 = 0$	$0 = a_1 + b_1 - 3C_1 + 1$ $0 = a_1 + b_1 + 1$ $a_1 + b_1 = -1$	$0 = -2b_1 - 1$ $2b_1 = -1$ $b_1 = -1/2$

a_1, b_1, c_1 sub in eq 1

$\pi_1 \quad = \quad H^{-1/2} g^{-1/2} \rho^0 V$

$$\pi_1 = \frac{v}{\sqrt{gh}}$$

Step 7

$\pi_2 = m^0 L^0 T^0 = \quad Hg\rho \quad D \quad = (L)^{a_2} (LT^{-2})^{b_2} (ML^{-3})^{c_2} \ L$

Equating power of 'm'	Equating power of 'L'	Equating power of 'T'
$0 = C_2$	$0 = a_2 + b_2 + 1$ $a_2 + b_2 = -1$ $a_2 + 0 = -1$ $a_2 = -1$	$0 = -2b_2$ $b_2 = 0$

a_2, b_2, c_2 sub in eqn. 2

$\pi_2 = H^{-1} g^0 \rho^0 D$

$$\boxed{\pi_2 = \frac{D}{H}}$$

Step 8

$\pi_3 = m^0 L^0 T^0 = Hg\rho \quad \mu = (L)^{a_3} (LT^{-2})^{b_3} (ML^{-3})^{c_3} \quad ML^{-1}T^{-1}$

Equating power of 'm'	Equating power of 'L'	Equating power of 'T'
$0 = c_3 + 1$ $c_3 = -1$	$0 = a_3 + b_3 - 3c_3 - 1$ $0 = a_3 - 1/2 - 3(-1) - 1$ $0 = a_3 + 3/2$ $a_3 = -3/2$	$0 = -2b_3 - 1$ $b_3 = -1/2$

a_3, b_3, c_3 sub in eq 3

$\pi_3 = H^{-3/2} g^{-1/2} \rho^{-1} \mu$

$$\boxed{\pi_3 = \frac{\mu}{H^{3/2} g^{-1/2} \rho^{1}}}$$

Step 9

$\pi_1 = \varphi [\pi_2, \pi_3]$

$$\frac{v}{\sqrt{gh}} = \varphi \left[\frac{D}{H}, \frac{\mu}{H^{3/2} g^{-1/2} \rho^{1}} \right]$$

From Pitot tube

$$V = \sqrt{2gh}$$

$$\frac{v}{\sqrt{2}} = \sqrt{gH}$$

$$\frac{v}{\sqrt{gh}} = \varphi \left\{ \frac{D}{H}, \frac{\mu}{H \frac{V}{\sqrt{2}} \rho} \right\}$$

$$\boxed{V = \sqrt{2gh}\, f \left(\frac{D}{H}, \frac{\mu}{\rho V H} \right)}$$

Hence Proved

Two Marks Question and Answers

1. Give the dimensions of the following physical quantities.

   a) Pressure (P) $\quad = \quad \dfrac{Force}{Area} = \left(\dfrac{MLT^{-2}}{L^2} \right) \quad = M\,L^{-1}\,T^{-2}$

   b) Surface tension (σ) $= \dfrac{N}{M} = \left(\dfrac{MLT^{-2}}{L} \right) = \quad MT^{-2}$

   c) Dynamic viscosity (μ) $\quad = \quad \dfrac{N-s}{m^2} = M\,L^{-1}\,T^{-1}$

   d) Kinematic viscosity (γ) $= \dfrac{m^2}{s} = \left(\dfrac{L^2}{T} \right) \quad = \quad L^2\,T^{-1}$

2. What is a dimensional homogeneous equation? Give example.

   It means, the dimensions of Right hand side and Left hand side are equal.

   Example $\quad$ V $\quad = \quad \sqrt{2gH}$

   m/s= $\quad \sqrt{2\dfrac{m}{s^2}\,xm}$

   m/s= $\quad \sqrt{2\dfrac{m^2}{s^2}}$

   L/T = $\quad \sqrt{\dfrac{L^2}{T^2}} \quad \dfrac{L}{T} = \dfrac{L}{T}$

3. State the Buckingham π theorem.

   It state that if there are 'n' variables in a dimensional homogeneous equation and if these variables contain 'm' fundamental dimensions (M, L, T), then they are grouped into n – m dimensionless independent π - term.

4. What are the similarities between model and prototype?

   1. Geometric similarity.
   2. Kinematic similarity.
   3. Dynamic similarity.

5. What is meant undistorted model?

   The model which is geometrically similar to its prototype is known as undistorted models. In such models the conditions of similitude are fully satisfied.

6.   State the methods of dimensional analysis.

    (i)   Rayleigh's method.

    (ii) Buckingham π - theorem.

7.   Describe briefly the selection of repeating variables in Buckingham π - theorem.

There is no separate rule for selecting repeating variables. But the number of repeating variables is equal to the fundamental dimensions of the problem. Generally, ρ, v, l, or ρ, v, D are choosen as repeating variables. It means, one refers to fluid property (ρ) one refers to flow property (v) and the other one refers to geometric property (l or D). In addition to this, the following points should be kept in mind while selecting repeating variables.

    (i)   No one variable should be dimensionless.

    (ii) The selected two repeating variables should not have the same dimensions.

    (iii) The selected repeating variables should be independent as far as possible.

8.   Define Weber number.

$$W_e = \sqrt{\frac{Inertia\ force}{Surface\ Tension\ force}}$$

9.   Define Reynolds number.

$$R_e = \sqrt{\frac{Inertia\ force}{Viscous\ force}}$$

10.   Define Mach number

$$M_e = \sqrt{\frac{Inertia\ force}{Elastic\ force}}$$

11.   Define Froude's Number

$$F_e = \sqrt{\frac{Inertia\ force}{Gravity\ force}}$$

12.   Define Euler's Number

$$E_u = \sqrt{\frac{Inertia\ force}{Pr\ essure\ force}}$$

13.   State the limitations of dimensional analysis.

    a.   Dimensional analysis does not give any clue regarding the selection of variables.

    b.   The complete information is not provided by dimensional analysis. It only indicates that there is some relationship between parameters.

c.   The values of co-efficient and the nature of function can be obtained only by experiments or from mathematical analysis.

14. Mention the applications of model testing.

   a.   Civil engineering structures such as dams, weirs, canals etc.

   b.   Design of harbour, ships and submarines.

   c.   Airplanes, rockets and machines, missiles.

15. In fluid flow, what does dynamic similarity mean? What are the non-dimensional numbers associated with dynamic similarity?

   a.   It is similarity of forces. The flows in the model and prototype are of dynamic similar.

   b.   Dimensional numbers are weight, force, dynamic viscosity, surface tension, capillarity etc.,

16. Mention the significance of Reynolds model law.

   a.   Motion of air planes.

   b.   Flow of incompressible fluid in closed pipes.

   c.   Motion of submarines, and

   d.   Flow around structures and other bodies immersed fully in moving fluids.

17. State Froude's model law.

Only gravitational force is more predetermining force. The law states, "the Froude number is same for both model and prototype". It is known as Froude model law.

18. Submarine is tested in the air tunnel. Identify the model law applicable.

Reynolds model law.

19. Mention the types of models

   a.   Undistorted models.

   b.   Distorted models.

20. What is meant by undistorted models?

The model which is geometrically similar to its prototype is known as undistorted models. In such models, the conditions of similitude are fully satisfied.

21. State three demerits of a distorted model.

   a.   Exit pressure and velocity distributions are not true.

   b.   A model wave may differ from that of prototype.

   c.   Both extrapolation and interpolation of results are difficult.

Unit III

Dimensional Analysis

Part-A

1. What do you understand by fundamental units and derived units?

2. What is Dimensionally Homogeneous equation and give an example?

3. State the advantages of Dimensional and model analysis.

4. State Buckingham's π theorem.

5. What is meant by dynamic similarity?

6. What is dynamic similarity?

7. Give the dimensions of the following Physical Quantities:

    a. Pressure

    b. Surface Tension

    c. Dynamic viscosity

    d. Kinematic Viscosity

8. What are the similarities between model and prototype?

9. In Fluid Flow, What does dynamic similarity mean? What are the non- dimensional numbers associated with dynamic similarity?

10. Submarine is tested in the air tunnel. Identify the model law applicable

11. Define and explain Reynolds number, Froude's number, Euler's number and Mach number.

12. Explain the different types of similarities that must exist between a prototype and its model.

13. Explain the different types of similarities exist between a prototype and its model.

14. State the Fourier law of dimensional homogeneity.

15. What are the uses of dimensional homogeneity?

16. What are the points to be remembered while deriving expressions using dimensional analysis?

17. State the methods of dimensional analysis.

18. How are the equations derived in Rayleigh's method?

19. Define Mach number.

20. State the Euler model law and give its significance.

Part - B

1. What are the criteria for selecting repeating variable in this dimensional analysis?

2. The resisting force(R) of a supersonic flight can be considered as dependent upon the length of the air craft 'l', velocity 'v' , air viscosity 'μ', air density 'ρ' and bulk modulus of air is 'k'. Express the functional relationship between these variables and the resisting force.

3. Using Buckingham's π theorem, show that velocity, through a circular pipe orifice is given by *H- head* causing flow; *D-dia* of orifice μ = Coefficient of viscosity p = mass density; g = acceleration due to gravity.

4. The efficiency (η of a fan depends on ρ (density), μ (viscosity) of the fluid, W (angular velocity), d (diameter of rotor) and Q (discharge). .Express η in terms of non-dimensional parameters. Use Buckingham's π theorem.

5. Using Buckingham's π- theorem, show that the velocity through a circular orifice in a pipe is given by $v = \sqrt{(2gH)}\, f\,\{d/\,H\,,\,\mu/\rho vH\}$ where v is the velocity through orifice of diameter d and H is the head causing the flow and ρ and μ are the density and dynamic viscosity of the fluid passing through the orifice and g is acceleration due to gravity.

6. Classify Models with scale ratios.

7. Write short notes on the following:

   a. Dimensionless Homogenity with example.

   b. Euler Model Law.

   c. Similitude.

   d. Undistorted and Distorted Models.

8. Explain Reynold's law of similitude and Froude's law of similitude.

9. The efficiency (η of a fan depends on ρ (density), μ (viscosity) of the fluid, W (angular velocity), d (diameter of rotor) and Q (discharge). Express η in terms of non-dimensional parameters. Use Buckingham's π theorem.

10. (i) Explain Reynold's law of similitude and Froude's law of similitude.

    (ii) Explain different types of similarities.

UNIT IV

PUMPS AND TURBINES

4.1. Turbines

Turbines are defined as the hydraulic machines which convert hydraulic energy into mechanical energy. This mechanical energy is used in running an electric generator which is directly coupled to the shaft of the turbine. Thus the mechanical energy is converted into electrical energy. The electric power which is obtained from the hydraulic energy (energy of water) is known as Hydro-electric power. At present the generation of hydro-electric power is the cheapest as compared by the power generated by other sources such as oil, coal, etc.

4.2. Difference between Turbine & Pump

Turbine	Pump
It is defined as hydraulic machines which converts hydraulic energy into mechanical energy, then mechanical energy is converted into electrical energy with help of generator	It is defined as hydraulic machines which converts electrical energy into mechanical energy, then mechanical energy is converted into hydraulic energy.

4.3. Difference between Impulse and Reaction Turbine

Impulse Turbine	Reaction Turbine
If at inlet of the turbine, the energy available is only Kinetic energy, the turbine is called as impulse turbine. Eg. Tangential flow Pelton Wheel turbine	If at inlet of the turbine, the water possess the kinetic energy as well as the pressure energy is known as Reaction Turbine. Eg. Kaplan Turbine, Francis Turbine

4.4. Efficiency of a Turbine

- Hydraulic efficiency (η_H)
- Mechanical efficiency (η_m)
- Volumetric efficiency (η_v)
- Overall efficiency (η_o)

4.4.1. *Hydraulic Efficiency (η_H)*

$$\eta_H = \frac{Runner\ Power(RP)}{Water\ Power(WP)}$$

$$\text{Water power (WP)} \quad = \quad \frac{\rho g Q H}{1000} \text{ kW}$$

Where,	ρ	=	Density of water	=	1000 kg/m^3
	Q	=	Discharge (m^3/s)		
	H	=	Total Head (m)		

4.4.2. Mechanical Efficiency (η_m)

$$\eta_m \quad = \quad \frac{Shaft\ Power(SP)}{Runner\ Power(RP)}$$

4.4.3. Volumetric Efficiency (η_v)

$$\eta_v \quad = \quad \frac{Volume\ of\ water\ Striking\ the\ runner}{Volume\ of\ water\ \text{sup}\ plied\ to\ the\ turbine}$$

4.4.4. Overall Efficiency (η_o)

$$\eta_o \quad = \quad \frac{Shaft\ Power(SP)}{Water\ Power(RP)} \quad \text{OR} \quad \eta_m \ x \ \eta_H$$

Shaft power → Pin kW

4.5.   Classification of Hydraulic Turbine

(1) According to the type of energy at inlet.

   a) Impulse Turbine.

   b) Reaction Turbine.

(2) According to the flow through runner.

   a)   Tangential flow turbine (Pelton Wheel).

   b)   Radial flow turbine (Francis Turbine).

   c)   Axial flow turbine (Kaplan Turbine).

   d)   Mixed flow turbine.

(3) According to the head at the inlet

   a)   High head turbine (Pelton wheel).

   b)   Medium head turbine (Francis Turbine).

   c)   Low Head turbine (Kaplan Turbine).

(4) According to the specific speed of the turbine.

   a)   Low specific speed turbine.

   b)   Medium specific speed turbine (Francis turbine).

   c)   High specific speed turbine (Kaplan Turbine).

4.6. Reciprocating Pump (Single Acting) (Positive Displacement Pump)

Synopsis

- Definition
- Diagram
- Parts

   a) Suction pipe

   b) Suction Valve

   c) Delivery Pipe

   d) Delivery Valve

   e) Piston & Cylinder

   f) Connecting rod

- Working
- Advantages

Definition

The Reciprocating pump is a positive displacement pump. It operates on the principle of actual displacement or pushing of liquid by a piston or plunger that reciprocates in a closely fitting cylinder.

Diagram

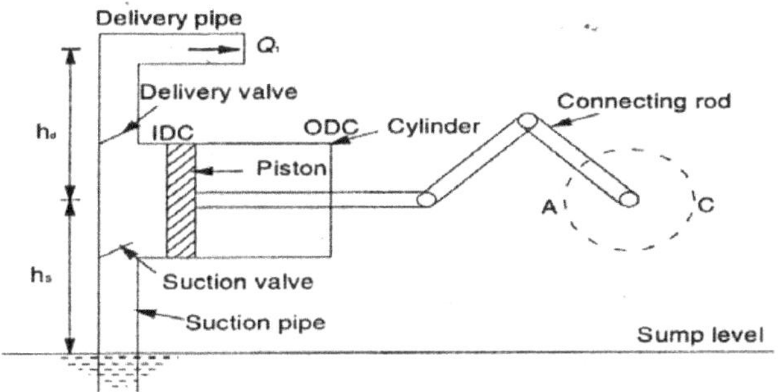

Figure 4.1: Single Acting Reciprocating Pump

Working

- Liquid acts on one side of the piston only.

- It has the suction pipe and delivery pipe placed above the sump.
- The crank rotates clockwise from IDC (Inner Dead Centre) to ODC.
- The piston moves outward to right and vaccum formed in left side.
- When the crank is at ODC, the suction stroke is completed and the cylinder is full of liquid.
- When the crank rotates from ODC to IDC, the piston moves inwards and a high pressure is build up in the cylinder.
- The increased pressure causes the discharged valve to open.
- Thus the liquid is carried out through the delivery pipe to the tank.

Advantages

- It is easy to operate.
- Cost is low.
- It is easily moveable.

4.7. Reciprocating Pump (Double Acting) (Positive Displacement Pump)

Synopsis

- Introduction
- Diagram
- Parts
  a) Suction head
  b) Delivery head
  c) Cylinder
  d) Piston
  e) Connecting rod
- Working
- Advantages

Definition

- It is a positive displacement pump.
- This pumps usually have one or more cylinders.

Diagram

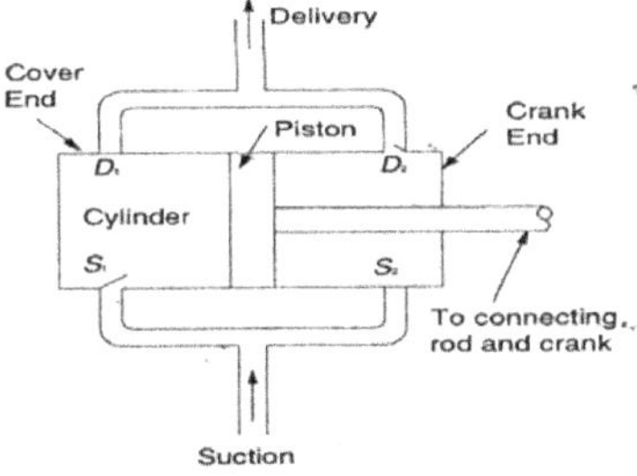

Figure 4.2: Double Acting Reciprocating Pump

Working

- In this process, suction and delivery stroke access simultaneously.
- When the crank rotates from IDC to ODC.
- Vaccum is created on it side of the piston. The liquid is sucked from the sump through suction valve.
- More uniform discharge because of continuous delivery strokes.

Advantages:

- Efficiency is high.
- Continuous Discharge.

4.8. Air Vessel Fitted on Reciprocating Pump

Synopsis

- Definition
- Diagram
- Parts
    a) Suction pipe
    b) Air vessel
    c) Cylinder
    d) Piston
    e) Connecting rod
- Working
- Advantages

Definition

It is a positive displacement pump mounted with Air Vessel. Air Vessel is a closed chamber made of cast iron. It is used to provide uniform discharge from reciprocating pump. It has opening at its base through which water flows into the air vessel or from the vessel depending on the quantity of water inside the cylinder. It is connected with both suction and delivery pipes separately. The water flows from the air vessel during the first half of the stroke and stored inside the vessel in the second half of the stroke by following the same procedure as that of delivery pipe. Due to these air vessels, the amount of work to be supplied to the reciprocating pump is reduced.

Diagram

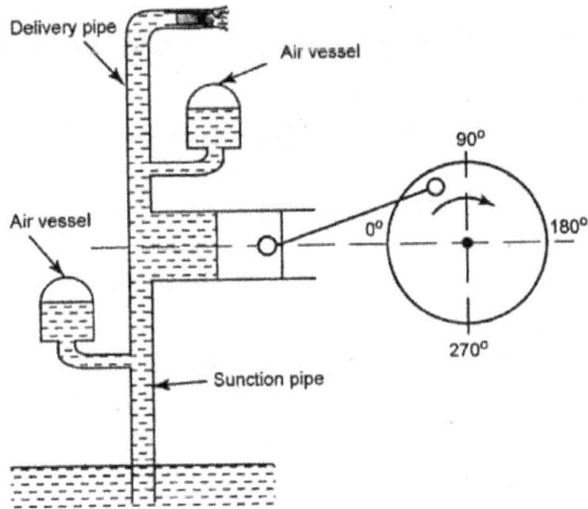

Figure 4.3: Air Vessel

Working

- Liquid acts on one side of the piston only.
- It has the suction pipe and delivery pipe placed above the sump.
- The crank rotates clockwise from IDC (Inner Dead Centre) to ODC.
- The piston moves outward to right and vaccum formed in left side.
- The water is forced at higher velocity than the mean velocity. Due to this, large quantity of water flows through the delivery pipe.
- Air vessel stores excess quantity of water during the first half of the delivery stroke and supplies excess quantity of water to the delivery pipe during the second half of the stroke.

- The water is stored in the air vessel by compressing the air.
- Similarly, the water is allowed to flow from the air vessel to the delivery pipe by expanding the compressed air.
- When the crank is at ODC, the suction stroke is completed and the cylinder is full of liquid.
- When the crank rotates from ODC to IDC, the piston moves inwards and a high pressure is build up in the cylinder.
- The increased pressure causes the discharged valve to open.
- Thus the liquid is carried out through the delivery pipe to the tank.

Advantages

- It is easy to operate
- Cost is low
- It is easily moveable
- Efficiency is high
- Continuous Discharge

4.9.   Centrifugal Pump (Rotary Pump)

Synopsis

<table>
<tr><td>

- Definition
- Diagram
- Parts
  a) Foot valve with strainer
  b) Suction pipe
  c) Delivery pipe
  d) Casing
  e) Impeller
  f) Priming valve
- Working
- Advantages

</td></tr>
</table>

Definition

- Centrifugal pumps are the devices which converts Mechanical Energy into Hydraulic Energy.

Diagram

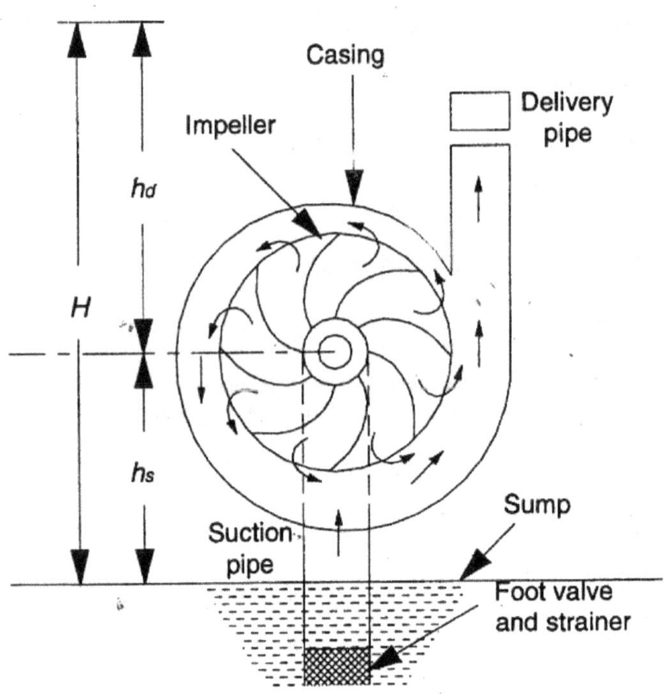

Figure 4.4: Centrifugal Pump

Working

- Delivery valve is closed
- The priming of pump is carried out. It involves filling the liquid in suction pipe and casing.
- The pump shaft and impellor is now rotated with the help of motor.
- The speed of the impeller should be sufficient to discharge from the delivery pipe.
- Now, delivery valve is opened. Liquid is lifted and discharge through the delivery pipe due to its' high pressure.

Advantages

- It is easily to operate.
- This pumps are used by the industrialist.

4.10. Pelton Wheel Turbine

Synopsis

- Definition
- Diagram
- Parts

  a) Spear

  b) Penstock

  c) Nozzle

  d) Runner

  e) Vanes (Bucket)

  f) Casing

  g) Brake jet

- Working
- Advantages

Definition

The pelton wheel is a tangential flow impulse turbine. The water strikes the bucket along the tangent of the runner. The energy available at the inlet of the turbine is only kinetic energy. The pressure at the inlet and outlet of the turbine is equal to atmospheric pressure.

Diagram

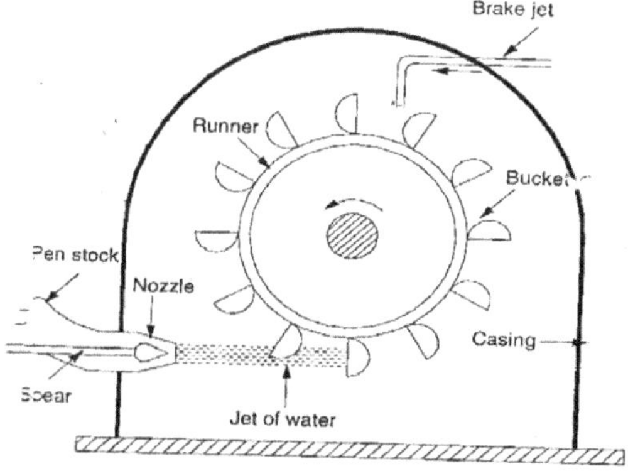

Figure 4.5: Pelton Wheel Turbine

Working

a) Runner with Brackets

- It consists of circular disc.
- The runner is mounted a horizontal shaft.
- It have a shape of double semi elliptical ridge.
- It shaped that the angle 10° to 20°.
- The water leaves its outer edge.

b) Outer Casing

- It made of cast iron.
- It has no hydraulic performs.
- It also acts a safe guard against accidents.

c) Brake Nozzle

- The nozzle is closed by moving spear in forward direction.
- The runner will revolve for long water due to inertia.
- The jet of water is called braking jet.

d) Governing Mechanism

- It used to regulate the water flow to the turbine.
- The quantity of water flowing through the runner in accordance any variation of load.

e) Pen Stock

- It is large size pipe. Conveys water from high level reservoir to the turbine.
- It is depend upon low head.
- The regulation of water flow the penstock is provided with control valves.

f) Spear and Nozzle

- It is fitted with an efficient nozzle.
- It is converts into hydraulic energy into kinetic energy.
- The water flows through the nozzle.
- The spear can move forward increasing the annular area.

g) Speed Ring

- The water passed through the speed ring.

- The number of stay vanes is usually taken as half number of guide vanes.
- It may be made of cast steel.

h) Guide Vanes

- The water passes through the series of guide vanes.
- The guide vanes direct the water onto the runner at appropriate design.
- The quantity of water supplied. The guide vanes are operated by wheel.

i) Runner

- It radial curved vanes are fixed.
- It made of cast steel.
- The runners force due to deviation effects.

j) Draft Tube

- It provides a negative suction heat at runner outlet by its possible to without any loss of head.

k) Advantages

- It is used in the high head and low discharge level.

4.11. Francis Turbine

Synopsis

<table>
<tr><td>

- Definition
- Diagram
- Parts

    a) Penstock

    b) Stay vane

    c) Guide vane

    d) Scroll casing

    e) Runner

    f) Draft tube

- Working
- Advantages

</td></tr>
</table>

Definition

- Inward flow reaction turbine.
- It had purely radial flow runner.

Diagram

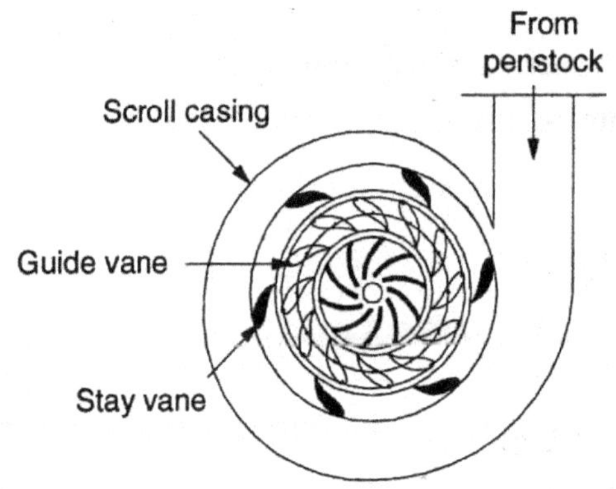

Figure 4.6: Francis Turbine

Working

a) Pen Stock

- It conveys water from the upstream of the dam to the turbine runner.

b) Scroll Casing

- It enters a scroll casing which completely surrounds the runner.
- It purpose of casing to provide an even distribution of water.
- It is depending upon the pressure.
- It can be rotated about pivots fixed to hub of the runner
- The water in the turbine turns through a right angle into the axial direction.

c) Stay Vane

From Scroll casing, the water passes through a stay vane. This consists of an upper and a lower ring held together by series of fixed vanes called Stay Vanes. The number of stay vanes is usually taken as half the number of guide vanes. The function of stay vane is to direct the water from the scroll casing to the guide vanes and also it resists the load imposed upon it. It may be made of cast iron / cast steel.

d) Guide vanes

From the speed ring, the water passes through a series of guide vanes. Thus, gates are provided all around the periphery of the turbine runner.

The guide vanes direct the water onto the runner at an appropriate angle as per the design. Also it is used to regulate the quantity of water supplied to the runner.

The guide vanes are airfoil shaped and they may be made of cast steel or plate steel.

e) Runner

It is a circular wheel on which a series of radial curved vanes are fixed. The vanes are so shaped that water enters the runner radially at outer periphery and leaves it axially at the inner periphery. The runners are made of cast steel, cast iron or stainless steel.

f) Draft Tube

After passing through the runner, the water is discharged to the tailrace through a gradually expanding tube called draft tube.

4.12. Kaplan Turbine

Synopsis

- Definition
- Diagram
- Parts
  a) Scroll Casing
  b) Stay ring
  c) Guide vanes
  d) Runner
  e) Draft Tube
- Working
- Advantages

Definition

- It is an axial flow reaction turbine.

- It requires large quantity of water to develop the power.

Diagram

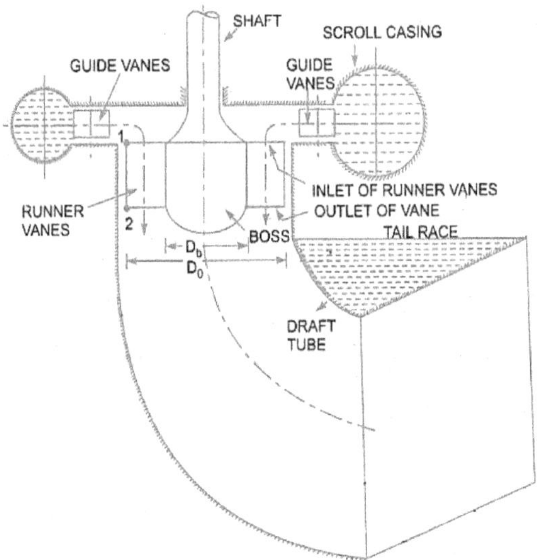

Figure 4.7: Kaplan Turbine

Working

a) Pen Stock

- It conveys water from the upstream of the dam to the turbine runner.

b) Scroll Casing

- It enters a scroll casing which completely surrounds the runner.
- It purpose of casing to provide an even distribution of water.
- It is depending upon the pressure.
- It can be rotated about pivots fixed to hub of the runner
- The water in the turbine turns through a right angle into the axial direction.

c) Stay Vane

From Scroll casing, the water passes through a stay vane. This consists of an upper and a lower ring held together by series of fixed vanes called Stay Vanes. The number of stay vanes is usually taken as half the number of guide vanes.

The function of stay vane is to direct the water from the scroll casing to the guide vanes and also it resists the load imposed upon it. It may be made of cast iron/cast steel.

d) Guide Vanes

From the speed ring, the water passes through a series of guide vanes. Thus, gates are provided all around the periphery of the turbine runner. The guide vanes direct the water onto the runner at an appropriate angle as per the design. Also it is used to regulate the quantity of water supplied to the runner. The guide vanes are airfoil shaped and they may be made of cast steel or plate steel.

e) Runner

It is a circular wheel on which a series of radial curved vanes are fixed. The vanes are so shaped that water enters the runner radially at outer periphery and leaves it axially at the inner periphery. The runners are made of cast steel, cast iron or stainless steel.

f) Draft Tube

After passing through the runner, the water is discharged to the tailrace through a gradually expanding tube called draft tube.

Advantages

- It is low head level turbine and discharge is high.

4.13. Vane Pump

Synopsis

- Definition
- Diagram
- Parts
  a) Inlet / Suction pipe
  b) Casing
  c) Vanes
  d) Rotating drum
  e) Receiver / Outlet pipe
- Working

Diagram

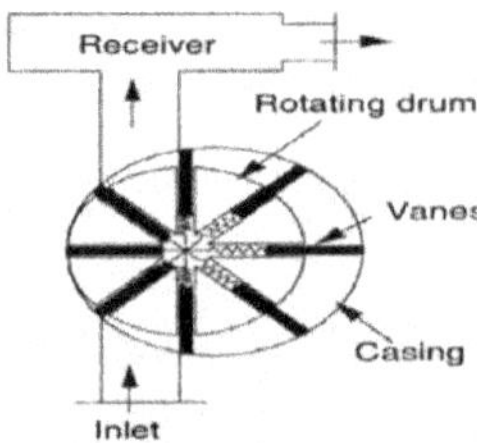

Figure 4.8: Vane Pump

Definition

- Vane pumps are also known as sliding vane.

Working

- The disc is eccentrically rotating inside the pump casing.
- Slots containing vanes are moved against the casing due to centrifugal force, when the disc rotates.
- Before starting the pump, it is filled with liquid.
- So, the liquid tight seal is formed.
- Due to continuous rotation of disc, the liquid is entrapped and forced to the delivery end with sufficient pressure.
- In some cases, springs are used to press the vanes against the casing
- But, vanes are generally hinged.

4.14. Lobe Pump

Synopsis

- Definition
- Diagram
- Part
  a) Inlet
  b) Outlet
  c) Lobe
  d) Casing
- Working
- Advantage & Disadvantage

Definition

- The action of lobe pump is similar to gear pump.
- Lobe pump is used to pump the Lubricant or high viscosity liquid.

Diagram

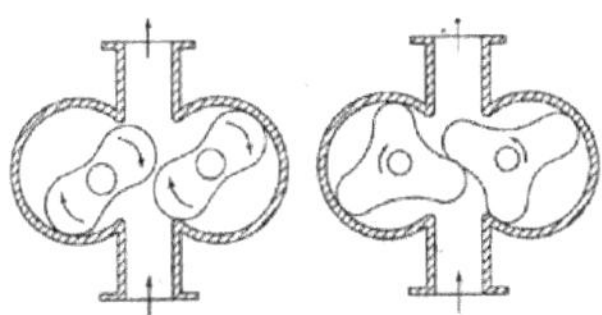

Figure 4.9: Lobe Pump

Working

- The specially designed wheels are arranged inside casing.
- The wheels have two or three lobes.
- The pump is filled with fluid before starting.
- The liquid makes a tight joint at the point of contact.
- The liquid is entrapped and flow to the delivery end when the lobes rotate.
- The pressure developed is due to back flow of fluid.

Advantage

- This pump is used to transfer the Lubricant or high viscosity liquid.

Disadvantage

- The main drawback is that the discharge is not uniform.

4.15. Gear Pump

Synopsis

<table>
<tr><td>

- Definition
- Part
    a) Suction
    b) Delivery
    c) Casing
    d) Gears
- Working

</td></tr>
</table>

Definition

It is defined as hydraulic machines which converts electrical energy into mechanical energy, then mechanical energy is converted into hydraulic energy. It is mounted with two gears. Gear pump is a positive displacement pump

Diagram

Figure 4.10: Gear Pump

Working

- A pair of gears would be rotating in mesh in a casing.
- One of the gear is connected to drive shaft and the other one is driven shaft.
- The pump is filled with liquid before starting.
- The liquid makes a tight joint at the point of contact.
- The liquid is entrapped and flow to the delivery end when the gears rotate.
- The pressure developed is due to force of fluid.
- So each and every tooth on the wheels acts as a piston of a reciprocating pump.

4.16. Screw Pump

Synopsis

- Definition
- Diagram
- Part
  a) Casing
  b) Flexible rotor
  c) Helical screw rotor
  d) Crank shaft
  e) Inlet
  f) Outlet
- Working
- Application

Definition

Screw pumps are rotary and positive displacement pumps that can have one or more screws to transfer high or low viscosity fluids along an axis.

Diagram

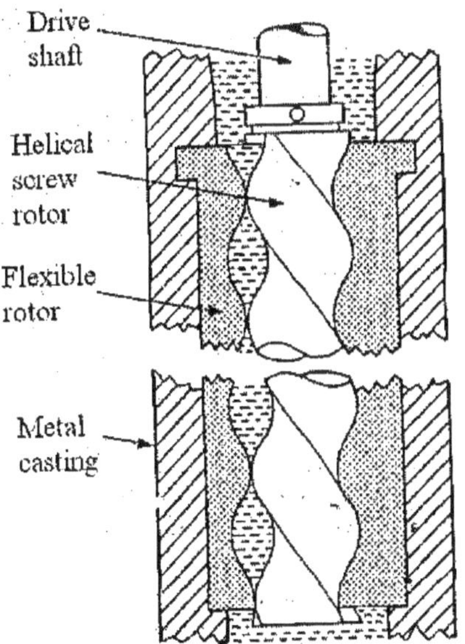

Figure 4.11: Screw Pump

Working

- Screw pump is rotary.
- Screw to transfer high or low viscosity fluid along an axis.
- Classical example, the screw pump is Archimedes screw pump.
- Screw pump having two or more intermeshing screw rotating axial clock wise or anticlockwise.
- Fluid is transferred to successive contact between the housing and screw flights from threads to next.

Application

- Use in high solid content, irrigation, agricultural system, and crude oil handling systems.

4.17. Required formula for Reciprocating Pump

1. Area of the piston $(A) = \dfrac{\pi}{4} D^2$ m²

    Where, D – Diameter of the piston (m)

2. Theoritical discharge for single acting reciprocating pump $(Q_t) = \dfrac{ALN}{60}$ (m³/s)

    Where, A – Area of the piston (m²)

    L – Length of the stroke (m)   (L = 2r => r = L/2)

    r – Radius of the crank (m)

    N – Speed of the pump (rpm)

3. Theoretical discharge for double acting reciprocating pum $(Q_t) = \dfrac{2ALN}{60}$ (m³/s)

4. Coefficient of discharge $(Cd) = \dfrac{Q_a}{Q_t}$

5. Actual discharge (Q_a) is given in the problems.

6. Slip $= Q_t - Q_a$

7. Percentage of slip $= \dfrac{Q_t - Q_a}{Q_t} x100$

8. Power required to drive the single acting pump $P = \dfrac{\rho g Q_t (h_s + h_d)}{60 x 1000}$ kW

    $= \dfrac{\rho g ALN (h_s + h_d)}{60 x 1000}$ kW

    Where, ρ = Density of water

    A = Area of the piston (m²

    L = Length of stroke = 2 r (m)

    N = Speed of the pump (rpm)

    h_s = Suction head (m)

    h_d = Delivery head (m)

9. Theoretical power required to drive double acting reciprocating pump

    $P = \dfrac{\rho g ALN (h_s + h_d)}{60 x 1000}$ kW

10. Acceleration head at the beginning of the delivery stroke $h_{ad} = \dfrac{ld}{g} \times \omega^2 r x Cos\,\theta x \dfrac{A}{ad}$

Where, ld = Length of delivery stroke (m)

Ω = Angular velocity (rad/s) = $\omega = \dfrac{2\pi N}{60}$ (rad / s)

r = Crank radius (m)

L = 2 r => r = L/2

L = Stroke length (m)

A = Area of the piston = $\dfrac{\pi}{4} D^2$ m²

D = Diameter of the piston (m)

ad = Area of the delivery pipe (m²) $= \dfrac{\pi}{4} d_d^{\ 2}$ .... m²

d_d = diameter of the delivery pipe (m)

Note: $\theta = 0$ at beginning of delivery stroke Cos 0 = 1.

$\theta = 90°$ at the middle of the delivery stroke Cos 90 = 0

Problems in Reciprocating Pump

1. A single acting reciprocating running at 50 rpm, delivers 0.01 m³/s of water. The diameter of the piston is 200 mm. The length of the stroke is 400 mm.

    (i) Theoretical discharge of pump

    (ii) Coefficient of discharge

    (iii) Slip

    (iv) Percentage of slip.

Given Data

N = 50 rpm

Q_a = 0.01 m³/s

D = 200 mm = 0.2 m

L = 400 mm = 0.4 m

Solution

Step 1

Theoretical discharge Q_t = $\dfrac{ALN}{60}$

$$= \frac{\pi}{4} x \frac{0.2^2 \, x0.4x50}{60}$$

$$= 0.01047 = 0.011 \text{ m}^3/\text{s}$$

Step 2

Coefficient of discharge => $Cd = \dfrac{Q_a}{Q_t} = \dfrac{0.01}{0.011}$ $=$ 0.955

Step 3

Slip $=$ $Q_t - Q_a$ $=$ $0.01047 - 0.01$ $=$ $0.00047 \text{ m}^3/\text{s}$

Step 4

Percentage of slip $=$ $\dfrac{Q_t - Q_a}{Q_t} x100$ $=$ $\dfrac{0.0047}{0.01047} x100 =$ 4.48%

2. A double acting reciprocating pump running at 400 rpm if discharge is 1 m³ of water per minute. A pump has a stroke of 400 mm, the diameter of the piston is 200 mm. The deliver and suction head are 20 m and 5 m respectively.

(i) Slip of the pump.

(ii) Theoretical discharge.

(iii) Power required to drive the pump.

Given Data

Double acting reciprocating pump

N	=	40 rpm			
Q_a	=	1 m³/min	=	(1/60) m³/s	=	0.0166 m³/s
L	=	400 mm= 0.4 m			
D	=	200 mm= 0.2 m			
h_s	=	20 m			
h_d	=	5 m			

Solution

Step 1

Theoretical discharge Q_t $=$ $\dfrac{2ALN}{60}$

$$= \quad \frac{2\pi}{4} \, x \, \frac{0.2^2 \, x40x0.4}{60}$$

$$= \quad 0.01674 \ \text{m}^3/\text{s}$$

Step 2

Slip	=	$Q_t - Q_a$	=	$0.01674 - 0.01666$
			=	0.000086 m³/s

Step 3

$$\text{Power required (P)} \quad = \quad \frac{2x\rho x g x ALNx(h_s + h_d)}{60x1000}$$

$$= \quad \frac{2x1000x9.81x2x\pi x0.2^2 \, x40x0.4x25}{60x1000}$$

$$P \quad = \quad 4.107 \ \text{kW}$$

3. The cylinder bore diameter of a single acting reciprocating pump is 150 mm and its stroke is 300 mm. The pump runs at 50 rpm and lifts water through the height of 25 m, the delivery pipe is 22 m long and 100 mm in diameter. Find the theoretical discharge and theoretical power required to run the pump. If actual discharge is 4.2 *l*/s. Find the percentage of slip. Also determine the acceleration head at the beginning and middle of delivery stroke.

Given Data: (Reciprocating Pump Single Acting)

D	=	150 mm	=	0.15 m
L	=	300 mm	=	0.3 m
N	=	50 rpm		
H	=	$h_s + h_d$	=	25 m
l_d	=	22 m		
d_d	=	100 mm	=	0.1 m
Q_a	=	4.2 lit / s	=	0.0042 m³/s.

Solution

Step 1

$$\text{Theoretical discharge} \quad Q_t = \frac{ALN}{60} = \frac{\pi}{4} \, x \, \frac{0.15^2 \, x0.3x50}{60}$$

$$= 0.00442 \ \text{m}^3/\text{s}$$

Step 2

$$\text{Theoretical Power (P)} = \frac{\rho \times g \times A L N x (h_s + h_d)}{60 \times 1000} = 9.81 x 25 x 0.0044 = 1.083 \text{ kW}$$

Step 3

$$\text{Percentage of slip} = \frac{Q_t - Q_a}{Q_t} x 100 = \frac{0.00442 - 0.0042}{0.00442} x 100 = 4.9\%$$

Step 4

Acceleration head in beginning ($\theta = 0$)

$$h_{ad} = \frac{ld}{g} x \omega^2 r x Cos\, \theta x \frac{A}{ad} \qquad \omega = \frac{2\pi N}{60} = 5.23 \text{ rad/s}$$

$$ad = \frac{\pi}{4} x d_d^2 = 0.785 x 0.1^2$$

$$ad = 0.00785 \text{ m}^2$$

$$r = L/2 = \frac{0.3}{2} = 0.15 \text{ m}$$

$$h_{ad} = \frac{22}{9.81} x 5.23^2 x 0.15 x Cos\; 0x \frac{0.0176}{0.00785} = 20.62 \text{ m}$$

Step 5

Acceleration head in middle ($\theta = 90°$)

$$h_{ad} = 0$$

4. The length and diameter of a suction pipe of a single acting reciprocating pump are 5 m and 10 cm respectively. The pump has a plunger of diameter 15 cm and a stroke length of 35 cm. The centre of the pump is 3 m above the water surface in the pump. The atmospheric pressure head is 10.3 m of water and pump is running at 35 rpm. Determine –

    (i) Pressure head due to acceleration at the beginning of the suction stroke.

    (ii) Maximum pressure head due to acceleration, and

    (iii) Pressure head in the cylinder at the beginning and at the end of the stroke.

Given

Length of suction pipe, (l_s)	=	5 m
Diameter of suction pipe (d_s)	=	10 cm = 0.1 m

| Area | (a_s) | $=$ | $\dfrac{\pi}{4}d_s^{\,2} = \dfrac{\pi}{4}x0.1^2$ | $=$ | 0.007854 m² |

Diameter of plunger,   D   =   15 cm = 0.15 m

Area of plunger   (A) $= \dfrac{\pi}{4}D^2 = \dfrac{\pi}{4}x0.15^2$

    =   0.01767 m²

Stroke length,   (L)   =   35 cm = 0.35 m

Crank radius,   r $= \dfrac{L}{2} = \dfrac{0.35}{2}$

    =   0.175 m

Suction head, h_s =   3 m

Atmospheric pressure head, H_{atm} =   10.3 m of water

Speed of pump,   N   =   35 rpm

Angular speed of the crank is given by,

$$\omega = \frac{2\pi N}{60} = \frac{2\pi x 35}{60} = 3.665 \text{ rad/s}$$

Solution

Step 1

The pressure head due to acceleration in the suction pipe is given by

$$h_{as} = \frac{l_s}{g}x\frac{A}{a_s}x\omega^2 r.\cos\theta$$

At the beginning of stroke, $\theta = 0$ and hence $\cos\theta = 1$

$$h_{as} = \frac{l_s}{g}x\frac{A}{a_s}x\omega^2 r.\cos\theta = \frac{5}{9.81}x\frac{0.01767}{0.007854}x3.665^2 x0.175$$

$$= 2.695 \text{ m.} \qquad \textbf{\textit{Ans.}}$$

Step 2

Maximum pressure head due to acceleration in suction pipe is given by

$$(h_{as})_{max} = \frac{l_s}{g}x\frac{A}{a_s}x\omega^2 r.\cos\theta$$

$$= 2.695 \text{ m. } \textbf{Ans.}$$

Step 3

Pressure head in the cylinder at the beginning of the suction stroke.

$$= \quad h_s + h_{as} \quad = \quad 3.0 + 2.695 \quad = \quad 5.695$$

This pressure head in the cylinder is below the atmospheric pressure head.

Absolute pressure head in the cylinder at the beginning of suction stroke.

$$= \quad \text{Atmospheric pressure head} - 5.695$$

$$= \quad 10.3 - 5.695 \quad = \quad 4.605 \text{ m of water (abs.)} \quad \textbf{Ans.}$$

Step 4

Similarly, pressure head in the cylinder at the end of suction stroke.

$$= \quad h_s - h_{as} \quad = \quad 3.0 - 2.695$$

$$= \quad 0.305 \text{ m below atmospheric pressure Head}$$

$$= \quad 10.3 - 0.305 \quad = \quad 9.995 \text{ m of water (abs.)} \quad \textbf{Ans.}$$

4.18. Euler's equation for Turbo Machines

The Euler's equation is expressed in terms of momentum equation for turbo machines.

Let, ρ -   Density of fluid

         a -   Area

         V_1 -   Absolute velocity

         V_{w_1} -   Whirl velocity at inlet

         V_{w_2} -   Whirl velocity at outlet

         u -   Blade velocity

So, the force exerted by the water in the direction of motion is given by

$$= \quad \rho \, a \, V_1 (V_{w_1} + V_{w_2})$$

But the momentum of the water or impulse   =   F x u

Momentum of the water   = $\rho \, a \, V_1 (V_{w_1} + V_{w_2}) x u$

This momentum or impulse is known as workdone by the fluid on the rotating element.

4.19. Impact of Jets

The liquid comes out in a jet from the outlet of a nozzle, which is fitted to a pipe through which the liquid is flowing under pressure. If some plate, which may be fixed or moving, is placed in the path of the jet, a force is exerted by the jet on the plate. This force is obtained from Newton's second law of motion or from impulse-momentum equation. Thus impact of jet means the force exerted by the jet on a plate which may be stationary or moving. The following cases of the impact of jet (i.e.) the force exerted by the jet on a plate, will be considered:

a) Force exerted by the jet on a stationary plate when
   i.   Plate is vertical to the jet
   ii.   Plate is inclined to the jet
   iii. Plate is curved.
b) Force exerted by the jet on a moving plate, when
   i.   Plate is vertical to the jet
   ii.   Plate is inclined to the jet, and
   ii.   Plate is curved.

4.20. Roto-Dynamic Machines

The device in which the fluid is in continuous motion and imparts energy conversion is known as fluid machines. It can also be stated that the machine is converting energy contained by a fluid into mechanical energy or mechanical into hydraulic energy by a roto dynamic element. The fluid machines are also called as turbo machines. The energy transfer takes place between fluid and roto dynamic element.

4.21. Derive the Expression for Pelton Wheel Work done by Water on the Runner

The shape of the vanes of the bucket on the Pelton wheel as shown in figure. The jet of water from the nozzle strikes the bucket at the splitter, which splits up the jet into two parts.

The splitter is the inlet tip and outer edge of the bucket is the outlet tip of the bucket. The inlet velocity triangle is drawn at the splitter and outlet velocity triangle is drawn at the outer edge of the bucket.

Let,

H   =   Net head acting on the Pelton wheel

            = H_g - h_f

Where,

H_g =   Gross head and h_f = $\dfrac{4\,fLV^2}{D * x2g}$

Where,

D*	=	Dia of Penstock
N	=	Speed of the wheel in rpm
D	=	Diameter of the wheel
d	=	diameter of the jet

Then,

V_1 = Velocity of jet at inlet = $\sqrt{2gH}$ .... (i)

$$u = u_1 = u_2 = \frac{\pi DN}{60}$$

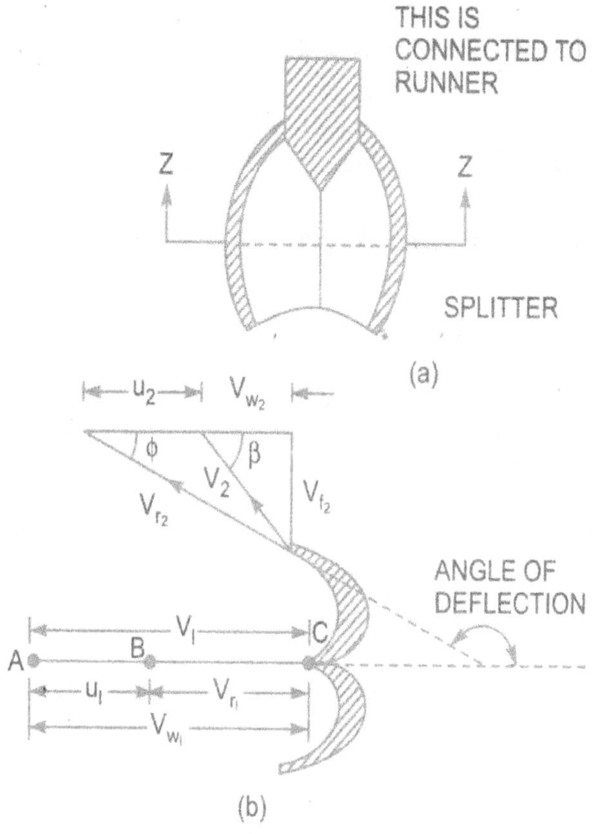

Figure 4.12

The velocity triangle at inlet will be a straight line where,

$$V_{r_1} \quad = \quad V_1 - u_1 = \quad V_1 - u$$

$$V_{w_1} \quad = \quad V_1$$

$$\propto \; = \; 0 \qquad \text{and } \theta = 0$$

From the velocity triangle at outlet, we have

$$V_{r_2} = V_{r_1} \qquad \text{and } V_{w_2} = V_{r_2} \cos \phi - u_2$$

The force exerted by the jet of water in the direction of motion is given by equation.

$$F_x \quad = \quad \rho \, a \, V_1 \left[V_{W_1} + V_{w_2} \right] \qquad \text{... (ii)}$$

As the angle β is an acute angle, +ve sign should be taken. Also this is the case of series of Vanes, the mass of water striking is $\rho a V_1$ and not $\rho a V_{r_1}$. In equation, 'a' is the area of the jet which is given as

$$a \quad = \quad \text{Area of jet} \quad = \quad \frac{\pi}{4} d^2$$

Now, work done by the jet on the runner per second.

$$= \quad F_x \, xu = \rho a V_1 [V_{W_1} + V_{w_2}] xu \quad \text{Nm/s} \qquad \text{... (iii)}$$

Power given to the runner by the jet

$$= \quad \frac{\rho a V_1 \left[V_{W_1} + V_{w_2} \right] xu}{1000} \quad \text{kW} \qquad \text{... (iv)}$$

Work done/s per unit weight of water striking

$$= \quad \frac{\rho a V_1 \left[V_{w_1} + V_{w_2} \right] xu}{\text{Weight of water striking}}$$

$$= \quad \frac{\rho a V_1 \left[V_{w_1} + V_{w_2} \right] xu}{\rho a V_1 xg} = \frac{1}{g} \left[V_{w_1} + V_{w_2} \right] xu \qquad \text{.. (v)}$$

The energy supplied to the jet at inlet is in the form of kinetic energy and is equal to $\frac{1}{2} mV^2$

$$\therefore \quad \text{K.E. of jet per second} \quad = \quad \tfrac{1}{2} (\rho a V_1) \times V_1^2$$

$$\therefore \quad \text{Hydraulic efficiency, } \eta_h \quad = \quad \frac{Work \; done \; per \; \sec ond}{K.E. of \; jet \; per \; \sec ond}$$

$$= \frac{\rho a V_1 \left[V_{w_1} + V_{w_2}\right] xu}{\frac{1}{2}(\rho a V_1) x V_1^2} = \frac{2\left[V_{w_1} + V_{w_2}\right] xu}{V_1^2} \quad .. \text{(vi)}$$

Now, $V_{w_1} = V_1$, $V_{r_1} = V_1 - u_1 = (V_1 - u)$

$V_{r_2} = (V_1 - u)$

And, $V_{w_2} = V_{r_2} Cos\,\phi - u_2 = V_{r_2} Cos\,\phi - u = (V_1 - u)Cos\,\phi - u$

Substituting the values of V_{w_1} and V_{w_2} in equation (vi)

$$\eta_h = \frac{2\left[V_1 + (V_1 - u)Cos\phi - u\right]xu}{V_1^2}$$

$$= \frac{2\left[V_1 - u + (V_1 - u)Cos\phi\right]xu}{V_1^2} = \frac{2(V_1 - u)[1 + Cos\phi]u}{V_1^2} \quad ... \text{(vii)}$$

The efficiency will be maximum for a given value of V_1 when

$$\frac{d}{du}(\eta_h) = 0 \quad \text{OR} \quad \frac{d}{du}\left[\frac{2u(V_1 - u)(1+\cos\phi)}{V_1^2}\right] = 0$$

$$\text{OR} \quad \frac{(1+\cos\phi)}{V_1^2}\frac{d}{du}(2uV_1 - 2u^2) = 0 \quad \text{OR} \quad \frac{d}{du}(2uV_1 - 2u^2) = 0 \quad \left(\frac{(1+\cos\phi)}{V_1^2} \neq 0\right)$$

$$\text{OR } 2V_1 - 4u = 0 \quad \text{OR} \quad u = \frac{V_1}{V_1^2} \quad ...\text{(viii)}$$

Equation (viii) states that hydraulic efficiency of a Pelton wheel will be maximum when the velocity of the wheel is half the velocity of the jet of water at inlet.

The expression for maximum efficiency will be obtained by substituting the value of $u = \dfrac{V_1}{2}$ in equation (vii)

$$\text{Max. } \eta_h = \frac{2\left(V_1 - \dfrac{V_1}{2}\right)(1+\cos\phi)x\dfrac{V_1}{2}}{V_1^2}$$

$$= \frac{2x\dfrac{V_1}{2}(1+\cos\phi)\dfrac{V_1}{2}}{V_1^2} = \frac{(1+\cos\phi)}{2}$$

S.no	Description	Pelton wheel turbine	Francis turbine	Kaplan turbine	Centrifugal pump
1.	TYPES	Tangential flow impulse turbine	Inward flow (radial section turbine)	Axial flow reaction turbine	Reverse of turbine(centrifugal action)
2.	VELOCITY TRIANGLE	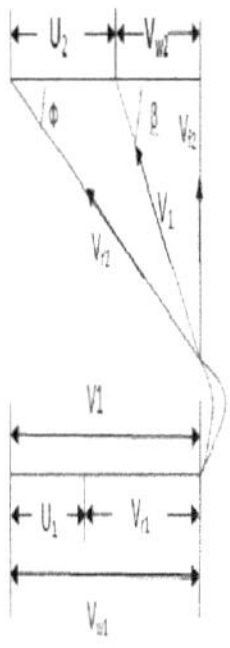			
3.	CONDITIONS	$\alpha = 0 \; ; \; \Theta = 0$ $V_{r1} = V_{r2}$ $U_1 = U_2 = U$ $\mathbf{V_1 = V_{w1}}$	$\beta = 90^0$ $V_{w2} = 0$ $V_2 = V_{f2}$	$\beta = 90^0$ $V_{w2} = 0$ $V_2 = V_{\Phi 2}$	
4.	INLET TRIANGLE	$\mathbf{V_1 = V_{w1}}$ $V_1 = V_{w1} - U_1$ $V_{r1} = V_1 - u_1$ $\mathbf{V_{r1} = V_1 - u}$	$\tan \Theta = \dfrac{V_{f1}}{V_{w1} - U_1}$ $\tan \alpha = \dfrac{V_{f1}}{V_{w1}}$	$\tan \Theta = \dfrac{V_{f1}}{V_{w1} - U_1}$ $\tan \alpha = \dfrac{V_{f1}}{V_{w1}}$	$\tan \Theta = \dfrac{V_1}{U_1} = \dfrac{V_{f1}}{U_1}$
5.	OUTLET TRIANGLE	$\cos\Phi = \dfrac{U_2 + V_{w2}}{V_{r2}}$ $V_{r2}\cos\Phi = U_2 + V_{w2}$ $V_{w2} = V_{r2}\cos\Phi - U_2$ $= V_{r1}\cos\Phi - U$ $\mathbf{V_{w2} = (v_1 - u)\cos\Phi - u}$	$\tan \Phi = \dfrac{V_{f2}}{U_2}$ $\tan \Phi = \dfrac{V_2}{U_2}$	$\tan \Phi = \dfrac{V_{f2}}{U_2}$ $\tan \Phi = \dfrac{V_2}{U_2}$	$\tan \Phi = \dfrac{V_{f2}}{U_2 - V_{w2}}$
6.	BUCKET SPEED (or) PERIPHERAL SPEED	$U_1 = U_2 = U = \dfrac{\pi DN}{60}\, m/s$ D = Bucket diameter (m) N = Bucket speed (rpm)	$U_1 \neq U_2$ $U_1 = \dfrac{\pi D_1 N}{60}$; $U_2 = \dfrac{\pi D_2 N}{60}$ D_1 = Inner diameter of impeller(m) D_2 = Outer diameter of impeller(m)	$U_1 = U_2 = U = \dfrac{\pi D_o N}{60}\,(m/s)$ D_o = Outer dia of the runner (m)	$U_1 \neq U_2$ $U_1 = \dfrac{\pi D_1 N}{60}$; $U_2 = \dfrac{\pi D_2 N}{60}$ D_1 = Inner diameter of impeller(m) D_2 = Outer diameter of impeller(m)
7.	BUCKET SPEED (or) PERIPHERAL SPEED	$U = \Phi\sqrt{2gH}$ (m/s) Φ = speed ratio (0.43 – 0.48)	$U = \Phi\sqrt{2gH}$ (m/s)	$U = \Phi\sqrt{2gH}$ (m/s)	--

S.no	Description	Pelton wheel turbine	Francis turbine	Kaplan turbine	Centrifugal pump
8.	AREA	Area of jet $a = \frac{\pi}{4}d^2$ (m²) d = diameter of jet	$A_1 = \pi D_1 B_1$ (m²) $A_2 = \pi D_2 B_2$ (m²) B_1 = Inner breadth of Impeller B_2 = Outer breadth of Impeller	$A = \frac{\pi}{4}(D_o^2 - D_b^2)$ (m²) D_o = Outer dia of the runner (m) D_b = Innerdia of the runner (m)	$A_1 = \pi D_1 B_1$ (m²) $A_2 = \pi D_2 B_2$ (m²) B_1 = Inner breadth of Impeller B_2 = Outer breadth of Impeller
9.	VELOCITY	$V_1 = c_v * \sqrt{2gH}$ (m/s) c_v = co efficient of velocity (0.98) H = Head (m)	$V_{f1} = V_{f2}$ $V_{f1} = V_{f2}$ = flow ratio* $\sqrt{2gH}$ (m/s)	$V_{f1} = V_{f2}$ = flow ratio* $\sqrt{2gH}$ (m/s)	$V_{f1} = V_{f2}$ (m/s)
10.	DISCHARGE	$q = Q = \frac{\pi}{4}d^2 * c_v * \sqrt{2gH}$ (m³/s)	$Q = \pi D_1 B_1 vf_1$ $Q = \pi D_2 B_2 vf_2$	$Q = \pi(D_o^2 - D_b^2)vf_1$ $Q = \pi(D_o^2 - D_b^2)vf_2$	$Q = \pi D_1 B_1 vf_1$ (m³/s) $Q = \pi D_2 B_2 vf_2$ (m³/s)
11.	OVERALL EFFICIENCY	$\eta_o = \dfrac{P}{\rho gQH/1000}*100$ P in KW $\eta_o = \dfrac{P*1000}{\rho gQH}*100$	$\eta_o = \dfrac{P}{\rho gQH/1000}*100$ P in KW $\eta_o = \dfrac{P*1000}{\rho gQH}*100$	$\eta_o = \dfrac{P}{\rho gQH/1000}*100$ P in KW $\eta_o = \dfrac{P*1000}{\rho gQH}*100$	$\eta_o = \dfrac{P}{\rho gQH/1000}*100$
12.	HYDRAULIC EFFICIENCY	$\eta_h = \dfrac{2(V_1-u)(1+\cos\Phi)(u)}{v_1^2}*100$ or $\eta_h = \dfrac{2(V_{w1}-V_{w2})(u)}{v_1^2}*100$	$\eta_h = \dfrac{V_{w1}U_1}{gH}*100$ $\eta_h = \dfrac{T.H - H.L}{T.H(INLET)}$		$\eta_{nano} = \dfrac{gH}{V_{w2}U_2}*100$
13.	WORK DONE	W.D = $\rho Q[V_{w1} - V_{w2}]*u$ W.D = $\dfrac{\rho Q[V_{w1}-V_{w2}]*u}{1000}$ (KW)	W.D = $\dfrac{1}{g}V_{w1}U_1$		W.D = $\dfrac{1}{g}V_{w2}U_2$
14.	POWER AVAILABLE AT NOZZLE	W.P = $\rho gQH/1000$ (Kw)			
15.	DESIGN	Design of Pelton Wheel 1)no of buckets $Z = 15 + \dfrac{D}{2d}$ D = Bucket diameter d = jet diameter 2) Width of the bucket W = 5*d (m) 3) Depth of the			

S.no	Description	Pelton wheel turbine	Francis turbine	Kaplan turbine	Centrifugal pump
		bucket D = 1.2d(m)			
16.	DEFLECTION ANGLE IS GIVEN	$\Phi = 180° -$ deflection angle			
17.	No of Jets	No of Jets $= \frac{Q}{q}$			
18.	Specific speed	$N = \frac{N\sqrt{P}}{H^{5/4}}$			$N = \frac{N\sqrt{Q}}{H_m^{3/4}}$

Problems in Pelton Wheel Turbine

5. A Pelton wheel is to be designed for a head of 60 m when running at 200 rpm. The Pelton wheel develops 95.6475 kW shaft power. The velocity of the buckets = 0.45 times the velocity of the jet, overall efficiency = 0.85 and coefficient of the velocity is equal to 0.98.

Given

Head	H	=	60 m
Speed	N	=	200 rpm
Shaft power,	S.P.	=	95.6475 kW
Velocity of bucket,	u	=	0.45 x velocity of jet
Overall efficiency	η_o	=	0.85
Coefficient of velocity,	C_v	=	0.98

Calculate

i. Design of Pelton wheel means to find diameter of jet (d),

ii. Diameter of wheel (D),

iii. Width and depth of buckets and

iv. Number of buckets on the wheel.

Solution

Step 1

(i) Velocity of jet, V_1

$$= C_v x \sqrt{2gH}$$

$$= 0.98x\sqrt{2 \ x \ 9.81 \ x \ 60} = \quad 33.62 \text{ m/s}$$

Bucket velocity, u

$$= u_1 = u_2$$

$$= 0.45 \ x \ V_1 = 0.45 \ x \ 33.62$$

$$= 15.13 \text{ m/s}$$

But u $= \dfrac{\pi DN}{60}$ Where D = Diameter of wheel

$$15.13 \quad = \quad \frac{\pi x D x 200}{60} \quad \text{or}$$

$$D = \frac{60 x 15.13}{\pi x 200} \quad = \quad 1.44 \text{ m} \quad \text{Ans.}$$

Step 2

(ii) Diameter of the jet (d)

Overall efficiency η_o = 0.85

But, η_o = $\dfrac{S.P.}{W.P.} = \dfrac{95.6475}{\left(\dfrac{W.P.}{1000}\right)} = \dfrac{95.6475}{\left(\dfrac{\rho x g x Q x H}{1000}\right)}$

$= \dfrac{95.6475 x 1000}{1000 x 9.81 x Q x 60}$

$Q \quad = \quad \dfrac{95.6475 x 1000}{\eta_o x 1000 x 9.81 x 60}$

$= \dfrac{95.6475 x 1000}{0.85 x 1000 x 9.81 x 60}$

$= \quad 0.1912 \text{ m}^3/\text{s}.$

But the discharge, Q = Area of jet x Velocity of jet

$0.1912 \quad = \quad \dfrac{\pi}{4} d^2 x V_1 = \dfrac{\pi}{4} d^2 x 33.62$

$d \quad = \quad \sqrt{\dfrac{4 x 0.1912}{\pi x 33.62}}$

$= 0.085 \text{ m} = 85 \text{mm}$

Step 3

Design of Pelton wheel

(iii) Size of buckets

Width of buckets = 5 x d = 5 x 85 = 425 mm

Depth of buckets = 1.2 x d = 1.2 x 85 = 102 mm **Ans.**

Step 4

(iv) Number of buckets on the wheel is given by

$Z \quad = \quad 15 + \dfrac{D}{2d} = 15 + \dfrac{1.44}{2 x 0.085} = 15 + 8.5$

$= \quad 23.5 \text{ say } 24 \text{ **Ans**}$

6. A Pelton wheel has a mean bucket speed of 10 metres per second with a jet of water flowing at the rate of 700 litres/s under a head of 30 metres. The buckets deflect the jet through an angle of 160°. Calculate the power given by water to the runner and the hydraulic efficiency of the turbine. Assume coefficient of velocity as 0.98.

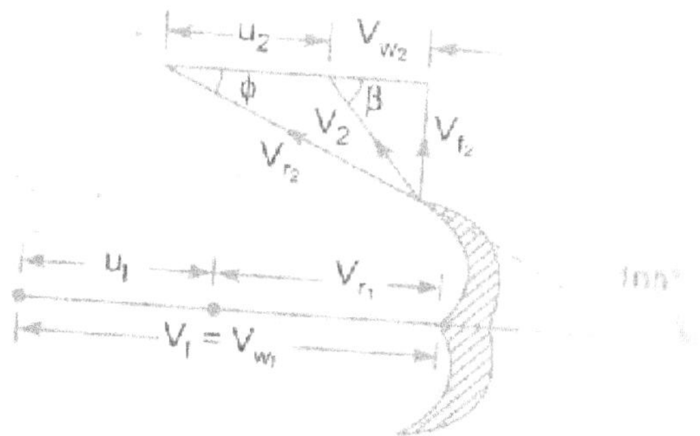

Figure 4.13

Given

Speed of bucket,	u	=	$u_1 = u_2 =$	10 m/s	
Discharge,	Q	=	700 litres/s = 0.7 m³/s,		
Head of water,	H	=	30 m		
Angle of deflection		=	160°		
Angle,	ϕ	=	180 – 160	= 20°	
Coefficient of velocity,	C_v	=	0.98		

Solution

Step 1

The velocity of jet,	V_1	=	$C_v x\sqrt{2gH} =$	$0.98x\sqrt{2 \ x \ 9.81 \ x \ 30}$
		=	23.77 m/s	
	V_{r_1}	=	$V_1 - u_1 =$	23.77 – 10
		=	13.77 m/s	
	V_{w_1}	=	V_1 =	23.77 m/s.

Step 2

From outlet velocity triangle,

$$V_{r_2} \quad = \quad V_{r_1} \quad = \quad 13.77 \text{ m/s}$$

$$V_{W_2} \quad = \quad V_{r_2} Cos\phi - u_2$$

$$= \quad 13.77 \cos 20 - 10.0 \quad = \quad 2.94 \text{ m/s.}$$

Step 3

Work done by the jet per second on the runner is given by

$$= \quad \rho \, x \, a \, x \, \left[V_{w_1} + V_{w_2} \right] x u$$

$$= \quad 1000 \times 0.7 \times [23.77 + 2.94] \times 10$$

$$(\because aV_1 = Q = 0.7 \text{ m}^3/\text{s})$$

$$= 186970 \text{ Nm/s.}$$

Step 4

Power given to turbine $\quad = \dfrac{186970}{1000} \quad = \quad 186.97 \text{ kW} \quad$ **Ans.**

Step 5

The hydraulic efficiency of the turbine is given by

$$\eta_h \quad = \quad \frac{2\left[V_{w_1} + V_{w_2} \right] x u}{V_1^{\,2}}$$

$$= \quad \frac{2[23.77 + 2.94 x 10]}{23.77 x 23.77}$$

$$= \quad 0.9454 \quad \text{or} \quad 94.54\% \text{ **Ans.**}$$

7. A Pelton wheel is to be designed for the following specifications:

Shaft power = 11,772 kW; Head = 380 metres; Speed = 750 rpm; Overall efficiency = 86%; Jet diameter is not to exceed one-sixth of the wheel diameter. Determine –

(i) The wheel diameter.

(ii) The number of jets required.

(iii) Diameter of the jet. Take $K_{v_1} = 0.985$ and $K_{u_1} \quad = \quad 0.45$

Given

Shaft power, $\qquad\qquad\qquad$ S.P. $\quad = \quad$ 11,772 kW

Head		H	=	380 m
Speed		N	=	750 rpm
Overall efficiency		η_o	=	86% or 0.86
Ratio of jet dia to wheel dia			=	$\dfrac{d}{D} = \dfrac{1}{6}$
Coefficient of velocity	$K_{v_1} = C_v$		=	0.985
Speed ratio	$K_{u_1} = 0.45$			

Solution

Step 1

Velocity of jet $\quad$ V_1 $\quad = \quad C_v x\sqrt{2gH}$

$$= \quad 0.985x\sqrt{2 \ x \ 9.81 \ x \ 380}$$

$$= \quad 85.05 \text{ m/s}$$

Step 2

The velocity of wheel, $\quad$ u $\quad = \quad u_1 = u_2$

$$= \quad \text{Speed ratio x } \sqrt{2gH}$$

$$= \quad 0.45x\sqrt{2 \ x \ 9.81 \ x \ 380}$$

$$= \quad 38.85 \text{ m/s}$$

But, $\quad$ u $\quad = \quad \dfrac{\pi DN}{60} \qquad \therefore 38.85 = \dfrac{\pi DN}{60}$

$$D \quad = \quad \dfrac{60x38.85}{\pi x N} = \dfrac{60x38.85}{\pi x 750}$$

$$= \quad 0.989 \text{ m } \textbf{Ans.}$$

$$\textbf{OR}$$

But $\quad \dfrac{d}{D} = \dfrac{1}{6}$

Step 3

Dia of jet $\quad$ d $\quad = \quad \dfrac{1}{6}xD = \dfrac{0.989}{6}$

$$= \quad 0.165 \text{ m } \textbf{Ans.}$$

Step 4

Discharge of one jet, q = Area of jet x Velocity of jet

$$= \frac{\pi}{4}d^2 x V_1 \qquad = \qquad \frac{\pi}{4}(0.165) x 85.05$$

$$= 1.818 \text{ m}^3/\text{s}. \qquad \dots \qquad (i)$$

Now, η_o =
$$\frac{S.P.}{W.P.} = \frac{11772}{\dfrac{\rho g x Q x H}{1000}}$$

$$0.86 = \frac{11772 x 1000}{1000 x 9.81 x Q x 380}$$

Where, Q = Total discharge.

Total discharge Q =
$$\frac{11772 x 1000}{1000 x 9.81 x 380 x 0.86} = 3.672 \text{ m}^3/\text{s}$$

Step 5

Number of jets =
$$\frac{Total\ discharge}{Discharge\ of\ one\ jet} = \frac{Q}{q} = \frac{3.672}{1.878}$$

$$= 2 \text{ jets } \textbf{Ans.}$$

8. The penstock supplies water from a reservoir to the Pelton wheel with a gross head of 500 m. One third of the gross head is lost in friction in the penstock. The rate of flow of water through the nozzle fitted at the end of the penstock is 2.0 m3/s. The angle of deflection of the jet is 165o. Determine the power given by the water to the runner and also hydraulic efficiency of the Pelton wheel. Take speed ratio = 0.45 and Cv = 1.0

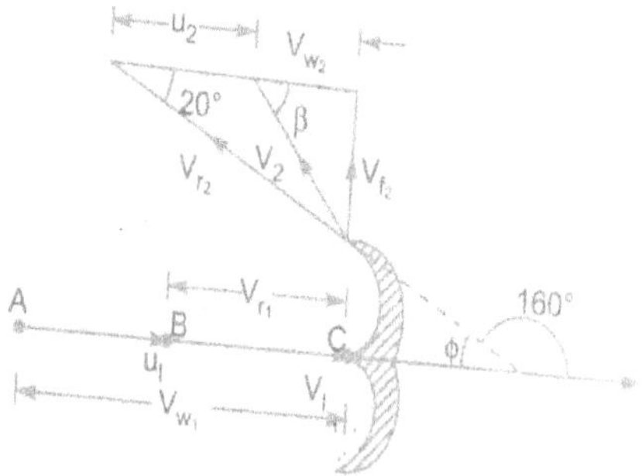

Figure 4.14

Given

Gross head	H_g	=	500 m
Head lost in friction,	h_f	=	$\dfrac{H_g}{3} = \dfrac{500}{3}$ = 166.7 m
Net head,	H	=	$H_g - h_f$ = 500 – 166.7 = 333.30 m
Discharge,	Q	=	2.0 m³/s.
Angle of deflection		=	165º
Angle	ϕ	=	180 – 165 = 15º
Speed ratio		=	0.45
Coefficient of velocity,	C_v	=	1.0

Solution

Step 1

$$\text{Velocity of jet,} \quad V_1 = C_v x \sqrt{2gH}$$

$$= 1.0 x \sqrt{2 \ x \ 9.81 \ x \ 333.3}$$

$$= 80.86 \text{ m/s}$$

Step 2

Velocity of wheel, $\quad$ u $\quad$ = $\quad$ Speed ratio x $\sqrt{2gH}$

OR $\quad u = u_1 = u_2 \quad = \quad 0.45 x \sqrt{2 \ x \ 9.81 \ x \ 333.3}$

$$= 36.387 \text{ m/s}$$

$$V_{r_1} = V_1 - u_1 = 80.86 - 36.87$$

$$= 44.473 \text{ m/s}$$

Also $\quad V_{w_1} = V_1 = 80.86$ m/s

Step 3

From outlet velocity triangle, we have

$$V_{r_2} = V_{r_1} = 44.473$$

$$V_{r_2} Cos\phi = u_2 + V_{w_2}$$

OR $\quad 44.473 \ Cos \ 15 = 36.387 + V_{w_2}$

OR
$$V_{w_2} = 44.473 \cos 15 - 36.387$$
$$= 6.57 \text{ m/s.}$$

Step 4

Work done by the jet on the runner per second is given by

$$\rho a V_1 \left[V_{w_1} + V_{w_2} \right] \times u = \rho Q \left[V_{w_1} + V_{w_2} \right] \times u \qquad (aV_1 = Q)$$
$$= 1000 \times 2.0 \times [80.86 + 6.57] \times 36.387$$
$$= 6362630 \text{ Nm/s}$$

Step 5

Power given by the water to the runner in kW

$$= \frac{Work\ done\ per\ \sec ond}{1000}$$
$$= \frac{6362630}{1000}$$
$$= 6362.63 \text{ kW} \qquad \textbf{Ans}$$

Step 6

Hydraulic efficiency of the turbine is given by

$$\eta_h = \frac{2\left[V_{w_1} + V_{w_2}\right] \times u}{V_1^2} = \frac{2[80.86 + 6.57] \times 36.387}{80.86 \times 80.86}$$
$$= 0.9731 \text{ or } 97.31\% \qquad \textbf{Ans}$$

9. A Pelton wheel is having a mean bucket diameter of 1 m and is running at 1000 rpm. The net head on the Pelton Wheel is 700 m. If the side clearance angle is 15° and discharge through nozzle is 0.1 m³/s.

Find

(i) Power available at the nozzle.

(ii) Hydraulic efficiency of the turbine.

Given

Diameter of wheel, D = 1.0 m

Speed of wheel, N = 1000 rpm

Tangential velocity of the wheel, u $= \dfrac{\pi DN}{60} = \dfrac{\pi \times 1.0 \times 1000}{60}$

		=	52.36 m/s
Net head on turbine	H	=	700 m
Side clearance angle,	ϕ	=	15°
Discharge,	Q	=	0.1 m³/s

Solution

Step 1

Velocity of jet at inlet V_1 = $C_v x \sqrt{2gH}$

$$= 1x\sqrt{2 \ x \ 9.81 \ x \ 700}$$

$$= 117.19 \text{ m/s}$$

(C$_v$ value is not given. Take it 1.0)

OR

Step 2

(i) Power available at the nozzle is given by

$$\text{W.P.} = \frac{WxH}{1000} = \frac{\rho xgxQxH}{1000}$$

$$= \frac{1000x9.81x0.1x700}{1000}$$

$$= 686.7 \text{ kW} \qquad \textbf{Ans.}$$

Step 3

(ii) Hydraulic efficiency is given by

$$\eta_h = \frac{2(V_1 - u)(1 + \cos\phi)u}{V_1^2}$$

$$= \frac{2(117.19 - 52.36)(1 + \cos 15)x52.36}{117.19^2}$$

$$= \frac{2x64.83x1.966x52.36}{117.19x117.19} = 0.9718$$

$$= 97.18\% \textbf{Ans.}$$

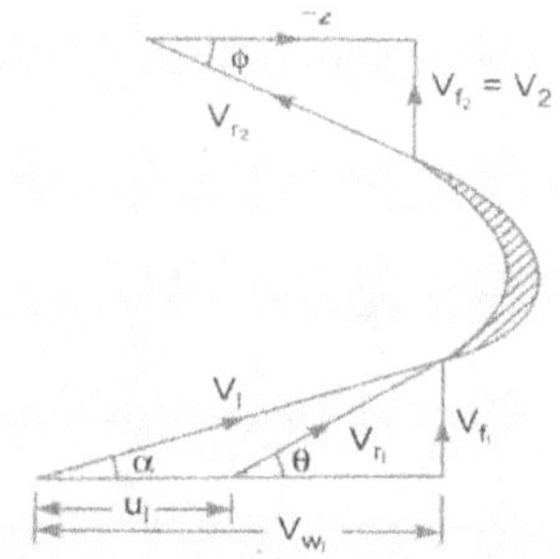

Figure 4.15

Problems in Francis Turbine

10. A Francis turbine with an overall efficiency of 75% is required to produce 148.25 kW power. It is working under a head of 7.62 m. The peripheral velocity = $0.26\sqrt{2gH}$ and the radial velocity of flow at inlet is $0.96\sqrt{2gH}$. The wheel runs at 150 rpm and the hydraulic losses in the turbine are 22% of the available energy. Assuming radial discharge, determine –

(i) The guide blade angle,

(ii) The wheel vane angle at inlet,

(iii) Diameter of the wheel at inlet, and

(iv) Width of the wheel at inlet.

Given

Overall efficiency	η_o	=	75%	=	0.75
Power produced	S.P.	=	148.25 kW		
Head	H	=	7.62 m		
Peripheral velocity,	u_1	=	$0.26\sqrt{2gH} = 0.26x\sqrt{2x9.81x7.62}$		
		=	3.179 m/s		
Velocity of flow at inlet	V_{f_1}	=	$0.96\sqrt{2gH} = 0.96x\sqrt{2x9.81x7.62}$		
		=	11.738 m/s		
Speed	N	=	150 rpm		
Hydraulic losses		=	22% of available energy		
Discharge at outlet		=	Radial		

$$V_{f_1} \quad = \quad 0 \text{ and } V_{f_2} \quad = \quad V_2$$

Solution

Step 1

Hydraulic efficiency is given as $\quad \eta_h \quad = \dfrac{Total\ head\ at\ inlet - Hydraulic\ loss}{Head\ at\ inlet}$

$$= \quad \frac{H - 0.22H}{H} = \frac{0.78H}{H}$$

$$= \quad 0.78$$

But $\qquad\qquad\qquad \eta_h \quad = \quad \dfrac{V_{w_1} u_1}{gH}$

$$0.78 \quad = \quad \frac{V_{w_1} u_1}{gH}$$

$$V_{w_1} \quad = \quad \frac{0.78 x g x H}{u_1} = \frac{0.78 x 9.81 x 7.62}{3.179}$$

$$= 18.34 \text{ m/s}$$

Step 2

(i) The guide blade angle, i.e., α. From inlet velocity triangle,

$$\tan \alpha \quad = \quad \frac{V_{f_1}}{V_{w_1}} = \frac{11.738}{18.34}$$

$$= \quad 0.64$$

$$\alpha \quad = \quad \tan^{-1}0.64 \quad = \quad 32.619° \text{ or } 32.37'. \qquad \textbf{Ans.}$$

Step 3

(ii) The wheel vane angle at inlet, i.e., θ

$$\tan \theta \quad = \quad \frac{V_{f_1}}{V_{w_1} - u_1} = \frac{11.738}{18.34 - 3.179} \quad = \quad 0.774$$

$$\theta = \tan^{-1}.774 = \quad 37.74 \quad \text{or } 37°44'. \qquad \textbf{Ans.}$$

Step 4

(iii) Diameter of wheel at inlet (D_1)

Using the relation, $\qquad u_1 \qquad = \qquad \dfrac{\pi D_1 N}{60}$

$$D_1 \quad = \quad \frac{60 x u_1}{\pi x N} = \frac{60 x 3.179}{\pi x 50} \quad = \quad 0.4047 \ \textbf{Ans.}$$

Step 5

(iv) Width of the wheel at inlet (B_1)

$$\eta o \quad = \quad \frac{S.P.}{W.P.} = \frac{148.25}{W.P.}$$

But, $\quad$ W.P. $\quad = \quad \dfrac{WH}{1000} = \dfrac{\rho x g x Q x H}{1000} = \dfrac{1000 x 9.81 x Q x 7.62}{1000}$

$$\eta o \quad = \quad \frac{148.25}{\dfrac{1000 x 9.81 x Q x 7.62}{1000}} = \frac{148.25 x 1000}{1000 x 9.81 x Q x 7.62}$$

OR

$$Q \quad = \quad \frac{148.25 x 1000}{1000 x 9.81 x 7.62 x \eta_o} = \frac{148.25 x 1000}{1000 x 9.81 x 0.75 x 7.62} = 2.644 \ m^3/s$$

Using equation $\quad Q \quad = \quad \pi D_1 \ x \ B_1 \ x \ V_{f_1}$

$$2.64 \quad = \quad \pi \ x \ 0.4047 \ x \ B_1 \ x \ 11.738$$

$$B_1 \quad = \quad \frac{2.644}{\pi x 0.4047 x 11.738} \quad = \quad 0.177 \ m \ \textbf{Ans.}$$

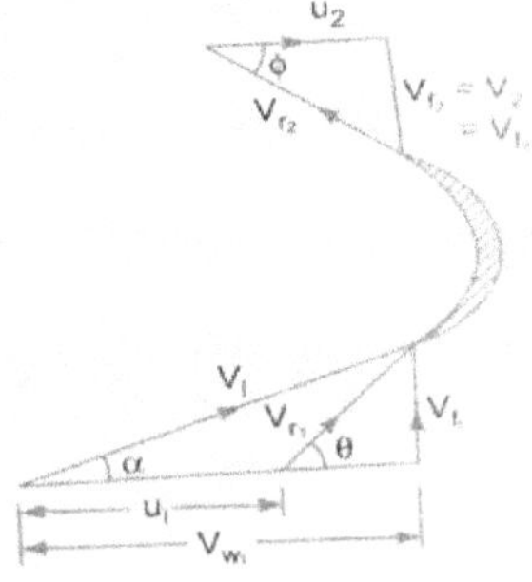

Figure 4.16

11. The following data is given for a Francis Turbine. Net head H = 60 m; Speed N = 700 rpm; shaft power = 294.3 kW; η_o = 84%; η_h = 93%; flow ratio =0.20; breadth ratio n = 0.1; Outer diameter of the runner = 2 x inner diameter of runner. The thickness of vanes occupy 5% of circumferential area of the runner, velocity of flow is constant at inlet and outlet and discharge is radial at outlet.

Determine:

   (i) Guide blade angle.

   (ii) Runner vane angles at inlet and outlet,

   (iii) Diameters of runner at inlet and outlet, and

   (iv) Width of wheel at inlet.

Given

Net head,	H	=	60 m,
Speed	N	=	700 rpm
Shaft power		=	294.3 kW
Overall efficiency,	η_o	=	84% = 0.84
Hydraulic efficiency	η_H	=	93% = 0.93

Solution

Step 1

Flow ratio, $\dfrac{V_{f_1}}{\sqrt{2gH}}$ = 0.20

$$V_{f_1} = 0.20 \times \sqrt{2gH}$$

$$= 0.20 \times \sqrt{2 \times 9.81 \times 60} = 6.862 \text{ m/s}$$

Breadth ratio, $\dfrac{B_1}{D_1}$ = 0.1

Outer diameter, D_1 = 2 x Inner diameter = 2 x D_2

Velocity of flow V_{f_1} = V_{f_2} = 6.862 m/s.

Thickness of vanes = 5% of circumferential area of runner.

Actual area of flow = 0.95 $\pi D_1 \times B_1$

Discharge at outlet = Radial

V_{w_2} = 0 and V_{f_2} = V_2

Using relation, η_o $=$ $\dfrac{S.P.}{W.P.}$

0.84 $=$ $\dfrac{294.3}{W.P.}$

W.P. $=$ $\dfrac{294.3}{0.84}$ $=$ $350.357\ kW$

But, W.P. $=$

$$\dfrac{WH}{1000} = \dfrac{\rho x g x Q x H}{1000} = \dfrac{1000 x 9.81 x Q x 60}{1000}$$

$\therefore$ $\dfrac{1000 x 9.81 x Q x 60}{1000}$ $=$ 350.357

$\therefore$ Q $=$ $\dfrac{35.357 x 1000}{60 x 1000 x 9.81}$

$=$ $0.5952\ m^3/s$

Using equation Q $=$ Actual area of flow x Velocity of flow

$=$ $0.95\ \pi D_1 \times B_1 \times V_{f_1}$

$=$ $0.95 \times \pi \times D_1 \times 0.1\ D_1 \times V_{f_2}$

$(\therefore B_1 = 0.1\ D_1)$

or 0.5952 $=$ $0.95 \times \pi \times D_1 \times 0.1 \times D_1 \times 6.862$

$=$ $2.048\ D_1^{\,2}$

$\therefore$ D_1 $=$ $\sqrt{\dfrac{0.5952}{2.048}}$ $=$ $0.54\ m$

But $\dfrac{B_1}{D_1}$ $=$ 0.1

$\therefore$ B_1 $=$ $0.1 \times D_1 =$ 0.1×0.54

$=$ $0.054\ m = 54\ mm$

Step 2

Tangential speed of the runner at inlet,

$$u_1 \quad = \quad \dfrac{\pi D_1 N}{60} = \dfrac{\pi x 0.54 x 700}{60} = 19.79\ m/s$$

Using relation for hydraulic efficiency,

$$\eta_h = \frac{V_{w_1} u_1}{gH} \quad or \quad 0.93 = \frac{V_{w_1} \times 19.79}{9.81 \times 60}$$

$$\therefore \quad V_{w_1} = \frac{0.93 \times 9.81 \times 60}{19.79} = 27.66 \ m/s.$$

Step 3

(i) Guide blade angle (α)

From inlet velocity triangle, $\tan \alpha = \dfrac{V_{f_1}}{V_{w_1}} = \dfrac{6.862}{27.66} = 0.248$

$$\therefore \quad \alpha = \tan^{-1} 0.248 = 13.928° \ or \ 13° \ 55.7'. \ \textbf{Ans.}$$

Step 4

(ii) Runner vane angles at inlet and outlet (θ and ϕ)

$$\tan \theta = \frac{V_{f_1}}{V_{w_1} - u_1} = \frac{6.862}{27.66 - 19.79} = 0.872$$

$$\theta = \tan^{-1} 0.872 = 41.09° \ or \ 41° \ 5.4' \quad \textbf{Ans.}$$

From outlet velocity triangle, $\tan \phi = \dfrac{V_{f_2}}{u_2} = \dfrac{V_{f_1}}{u_2} = \dfrac{6.862}{u_2}$ ... (i)

But $\quad u_2 = \dfrac{\pi D_2 N}{60} = \dfrac{\pi \times D_1}{2} \times \dfrac{N}{60} \ \left(\therefore D_2 = \dfrac{D_1}{2} given \right)$

$$= \pi \times \frac{0.54}{2} \times \frac{700}{60} = 9.896 \quad m/s$$

Substituting the value of u_2 in equation (i)

$$\tan \phi = \frac{6.862}{9.896} = 0.6934$$

$$\therefore \ \theta = \tan^{-1} 0.6934° = 34.74 \ or \ 34° \ 44.4' \qquad \textbf{Ans.}$$

Step 5

(iii) Diameters of runner at inlet and outlet

$D_1 = 0.54 \ m, \ D_2 = 0.27 \ m \quad \textbf{Ans.}$

Step 6

(iv) Width of wheel at inlet

$B_1 = 54 \ mm. \quad \textbf{Ans.}$

Problems in Kaplan Turbine

12. A Kaplan turbine working under a head of 20 m develops 11772 kW shaft power. The outer diameter of the runner is 3.5 m and hub diameter 1.75 m. The guide blade angle at the extreme edge of the runner is 35º. The hydraulic and overall efficiencies of the turbines are 88% and 84% respectively. If the velocity of whirl is zero at outlet, determine –

(i)   Runner vane angles at inlet and outlet at the extreme edge of the runner,

(ii) Speed of the turbine.

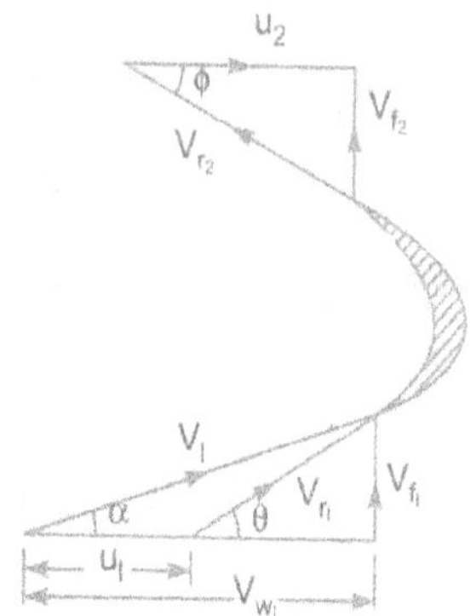

Given

Head,	H	=	20 m
Shaft power	S.P.	=	11772 kW
Outer dia of runner	D_o	=	3.5 m
Hub diameter,	D_b	=	1.75 m
Guide blade angle,	α	=	35º
Hydraulic efficiency	η_h	=	88%
Overall efficiency	η_o	=	84%
Velocity of whirl at outlet		=	0

Solution

Step 1

Using the relation, $\eta_o \quad = \quad \dfrac{S.P.}{W.P.}$

Where, W.P. $= \dfrac{W.P.}{1000} = \dfrac{\rho x g x Q x H}{1000}$, we get

$$0.84 \quad = \quad \frac{11772}{\dfrac{\rho x g x Q x H}{1000}} = \frac{11772 x 1000}{1000 x 9.81 x Q x 20}$$

$$(\because \rho = 1000)$$

$$\therefore \quad Q \quad = \quad \frac{11772 x 1000}{0.84 x 1000 x 9.81 x 20} \quad = \quad 71.428 \ \text{m}^3/\text{s}.$$

Using equation $\quad Q \quad = \quad \dfrac{\pi}{4}\left(D_o^{\,2} - D_b^{\,2}\right) x V_{f_1}$

Step 2

$$\text{OR} \quad 71.428 = \quad \frac{\pi}{4}\left(3.5^2 - 1.75^2\right) x V_{f_1} = \frac{\pi}{4}(12.25 - 3.0625)V_{f_1}$$

$$= \quad 7.216 \ V_{f_1}$$

$$V_{f_1} \quad = \quad \frac{71.428}{7.216} = 9.9 \ \ \text{m/s}$$

From inlet velocity triangle, $\tan \alpha = \dfrac{V_{f_2}}{V_{w_1}}$

$$V_{w_1} \quad = \quad \frac{V_{f_1}}{\tan \ \alpha} = \frac{9.9}{\tan \ 35} = \frac{9.9}{0.7} = 14.14 \qquad \text{m/s}$$

Step 3

Using the relation for hydraulic efficiency,

$$\eta_h \quad = \quad \frac{V_{w_1} u_1}{gH} \qquad\qquad \because \left(V_{w_2} = 0\right)$$

$$0.88 \quad = \quad \frac{14.14 x u_1}{9.81 x 20}$$

$$u_1 \quad = \quad \frac{0.88 x 9.81 x 20}{14.14} = \quad 12.21 \ \ \text{m/s}$$

Step 4

(i) Runner vane angles at inlet and outlet at the extreme edge of the runner are given as:

$$\tan \theta = \quad \frac{V_{f_1}}{V_{w_1} - u_1} = \frac{9.9}{(14.14 - 12.21)} \quad = \quad 5.13$$

$$\theta \quad = \quad \tan^{-1} 5.13 = 78.97° \ \text{OR} \ 78° \ 58' \quad \textbf{Ans}$$

For Kaplan turbine, $\quad$ $u_1 = u_2 = 12.21$ m/s and $V_{f_1} = V_{f_2} =$ $\quad$ 9.9 m/s

From outlet velocity triangle, $\tan\phi$ $\quad = \quad \dfrac{V_{f_2}}{u_2} = \dfrac{9.9}{12.21}$ $\quad = \quad$ 0.811

$$\phi \quad = \quad \tan^{-1} 0.811 = 39.035° \text{ or } 39° 2'. \quad \textbf{Ans.}$$

Step 5

(ii) Speed of turbine is given by $u_1 = u_2$ $\quad = \quad \dfrac{\pi D_o N}{60}$

$$12.21 \quad = \quad \dfrac{\pi x 3.5 x N}{60}$$

$$N \quad = \quad \dfrac{60 x 12.21}{\pi x 3.50} = 66.63 \text{ rpm} \quad \textbf{Ans.}$$

13. A Kaplan turbine develops 24647.6 kW power at an average head of 39 metres, assuming a speed ratio of 2, flow ratio of 0.6, diameter of the boss equal to 0.35 times the diameter of the runner and an overall efficiency of 90%, calculate the diameter, speed and specific speed of the turbine.

Given

Shaft power,	S.P.	=	24647.6 kW
Head	H	=	39 m

Step 1

Speed ratio, $\quad u_1 \sqrt{2gH}$ $\quad = \quad$ 2.0

$$u_1 \quad = \quad 2.0 \times \sqrt{2gH} = 2.0 \times \sqrt{2 x 9.81 x 39}$$

$$= \quad 55.32 \text{ m/s}$$

Step 2

Flow ratio, $\quad \dfrac{V_{f_1}}{\sqrt{2gH}} = $ $\quad$ 0.6

$$\therefore \quad V_{f_1} \quad = \quad 0.6 \times \sqrt{2gH} = 0.6 \times \sqrt{2 x 9.81 x 39} = 16.59 \text{ m/s}$$

Step 3

Diameter of boss $\quad = \quad$ 0.35 x Diameter of runner

$$D_b \quad = \quad 0.35 \times D_o$$

Step 4

| Overall efficiency η_o | = | 90% | = | 0.90 |

Using the relation, $\eta_o = \dfrac{S.P.}{W.P.}$, where $W.P. = \dfrac{\rho \times g \times Q \times H}{1000}$

$$0.90 = \frac{24647.6}{\dfrac{\rho \times g \times Q \times H}{1000}} = \frac{24647.6 \times 1000}{1000 \times 9.81 \times Q \times 39}$$

$$Q = \frac{24647.6 \times 1000}{1000 \times 9.81 \times 39 \times 0.9}$$

$$= 71.58 \text{ m}^3/\text{s}.$$

But from equation, we have

$$Q = \frac{\pi}{4}\left(D_o^2 - D_b^2\right) \times V_{f_1}$$

$$\left(\because D_b = 0.35 D_o, \quad V_{f_1} = 16.59\right)$$

$$71.58 = \frac{\pi}{4}\left(D_o^2 - (0.35 D_o)^2\right) \times 16.59$$

$$71.58 = \frac{\pi}{4}\left(D_o^2 - 0.1225 D_o^2\right) \times 16.59$$

$$71.58 = \frac{\pi}{4} \times 0.8775 D_o^2 \times 16.59 = 11.433\, D_o^2$$

Step 5

$$\text{(i)} \quad D_o = \sqrt{\frac{71.58}{11.433}}$$

$$= 2.5 \text{ m} \qquad \textbf{Ans.}$$

$$D_o = 0.35 \times D_o = 0.35 \times 2.5$$

$$= 0.875\text{m } \textbf{Ans.}$$

Step 6

(ii) Speed of the turbine is given by $u_1 = \dfrac{\pi D_o N}{60}$

$$55.32 = \frac{\pi \times 2.5 \times N}{60}$$

$$N \quad = \quad \frac{60 x 55.32}{\pi x 2.5} \quad = 422.61 \text{ rpm.} \quad \textbf{Ans.}$$

Step 7

(iii) Specific speed given by N_s $\quad = \quad \dfrac{N\sqrt{P}}{H^{5/4}}$ where P = Shaft power in kW

$$N_s \quad = \quad \frac{422.61 x \sqrt{24647.6}}{(39)^{5/4}} = \frac{422.61 x 156.99}{97.461}$$

$$= 680.76 \text{ rpm.} \quad \textbf{Ans.}$$

14. A Kaplan turbine runner is to be designed to develop 9100 kW. The net available head is 5.6 m. If the speed ratio = 2.09, flow ratio = 0.68; overall efficiency 86%, and the diameter of the boss is 1/3 the diameter of the runner, Find the diameter of the runner, its speed and the specific speed of the turbine.

Given

Power,	P	=	9100 kW		
Net Head	H	=	5.6 m		
Speed ratio,		=	2.09		
Flow ratio,		=	0.68		
Overall efficiency	η_o	=	86%	=	0.86
Diameter of boss		=	(1/3)	of diameter of runner	
	D_b	=	(1/3) D_o		

Step 1

Now Speed ratio $\quad = \quad \dfrac{u_1}{\sqrt{2gH}}$

$$u_1 \quad = \quad 2.09 \times \sqrt{2 x 9.81 x 5.6} \quad = \quad 21.95 \quad \text{m/s}$$

Step 2

Flow ratio $\quad = \quad \dfrac{V_{f_1}}{\sqrt{2gH}}$

$$V_{f_1} \quad = \quad 0.68 \times \sqrt{2 x 9.81 x 5.6}$$

$$= \quad 7.12 \text{ m/s}$$

Step 3

The overall efficiency is given by, $\eta_o = \dfrac{P}{\dfrac{\rho x g x Q x H}{1000}}$

OR $\quad Q = \dfrac{P x 1000}{\rho x g x H x \eta_o} = \dfrac{9100 x 1000}{1000 x 9.81 x 5.6 x 0.86}$

$$(\because \rho g = 1000 x 9.81 N/m^3)$$

$$= 192.5 \ m^3/s$$

Step 4

The discharge through a Kaplan Turbine is given by

$$Q = \frac{\pi}{4}\left[D_o^{\ 2} - D_b^{\ 2}\right] x V_{f_1}$$

$$192.5 = \frac{\pi}{4}\left[D_o^{\ 2} - \left(\frac{D_o}{3}\right)^2\right] x 7.12$$

$$192.5 = \frac{\pi}{4}\left[1 - \frac{1}{9}\right]D_o^{\ 2} x 7.12$$

$$D_o = \sqrt{\frac{4 x 192.5 x 9}{\pi x 8 x 7.12}}$$

$$= 6.21 \ m \qquad \textbf{Ans}$$

Step 5

The speed of turbine is given by, $u_1 = \dfrac{\pi D N}{60}$

$$N = \frac{60 x u_1}{\pi x D} = \frac{60 x 21.95}{\pi x 6.21}$$

$$= 67.5 \ rpm \qquad \textbf{Ans.}$$

Step 6

The specific speed is given by

$$N_s = \frac{N\sqrt{P}}{H^{5/4}} = \frac{67.5 x \sqrt{9100}}{5.6^{5/4}}$$

$$= 746 \qquad \textbf{Ans.}$$

15. A Kaplan turbine runner is to be designed to develop 7357.5 kW shaft power. The net available head is 5.50 m. Assume that the speed ratio is 2.09 and flow ratio is 0.68, and the overall efficiency is 60%. The diameter of the boss is 1/3rd of the diameter of the runner. Find the diameter of the runner, its speed and its specific speed.

Given

Shaft power,	P	=	7357.5 kW
Head	H	=	5.50 m

Step 1

$$\text{Speed ratio} \quad = \quad \frac{u_1}{\sqrt{2gH}} \quad = \quad 2.09$$

$$u_1 \quad = \quad 2.09 \times \sqrt{2x9.81x5.50} \quad = \quad 21.71 \quad m/s$$

Step 2

$$\text{Flow ratio} \quad = \quad \frac{V_{f_1}}{\sqrt{2gH}} \quad = \quad 0.68$$

$$V_{f_1} \quad = \quad 2.68 \times \sqrt{2x9.81x5.50} \quad = \quad 7.064 \quad m/s$$

Step 3

$$\text{Overall efficiency} \quad \eta_o \quad = \quad 60\% \quad = \quad 0.6$$

Step 4

$$\text{Diameter of boss,} \quad D_b \quad = \quad (1/3) \times D_o$$

$$\text{Using relation,} \quad \eta_o \quad = \quad \frac{Shaft\ power}{Water\ power} = \frac{7357.5}{\dfrac{\rho x g x Q x H}{1000}}$$

$$0.6 \quad = \quad \frac{7357.5x1000}{\rho x g x Q x H} = \frac{7357.5x1000}{1000x9.81xQx5.5}$$

$$Q \quad = \quad \frac{7357.5x1000}{1000x9.81x0.6x5.5} \quad = \quad 227.27 \quad m^3/s.$$

Step 5

Using equation for discharge

$$Q \quad = \quad \frac{\pi}{4}\left(D_o^{\ 2} - D_b^{\ 2}\right)xV_{f_1}$$

$$\left(\because D_b = \frac{D_o}{3}\right)$$

$$227.27 \quad = \quad \frac{\pi}{4}\left[D_o{}^2 - \left(\frac{D_o}{3}\right)^2\right]x7.064$$

$$= \quad \frac{\pi}{4}x\frac{8}{9}D_o{}^2 x7.064 = 4.9316 D_o{}^2$$

$$D_o \quad = \quad \sqrt{\frac{227.27}{4.9316}} \quad = \quad 6.788 \text{ m } \textbf{Ans.}$$

And $\quad$ $D_b \quad = \quad \dfrac{1}{3}x6.788 \quad = \quad 2.262 \quad$ m $\qquad$ **Ans.**

Using the relation, $u_2 \quad = \quad \dfrac{\pi D_o xN}{60}$

$$N \quad = \quad \frac{60xu_1}{\pi D_o} = \frac{60x21.71}{\pi x6.788}$$

$$= \quad 61.08 \text{ rpm} \qquad \textbf{Ans.}$$

Step 6

The specific speed (N_s) is given by

$$N_s \quad = \quad \frac{N\sqrt{P}}{H^{5/4}} = \frac{61.08x\sqrt{7357.5}}{5.50^{5/4}}$$

$$= \quad 622 \text{ rpm} \qquad \textbf{Ans.}$$

4.22. Derive the Expression for Work Done by the Centrifugal Pump {(or) By Impeller} on Water

The expression for the work done by the impeller on the water is obtained by drawing velocity triangles at inlet and outlet of the impeller in the same way as for a turbine. The water enters the impeller radially at inlet for best efficiency of the pump, which means the absolute velocity of water at inlet makes an angle of 90° with the direction of motion of the impeller at inlet. Hence angle $\propto$ = 90° and $V_{w_1} = 0$. For drawing the velocity triangles, the same notations are used as that for turbines. Figure shows the velocity triangles at the inlet and outlet tips of the vanes fixed to an impeller.

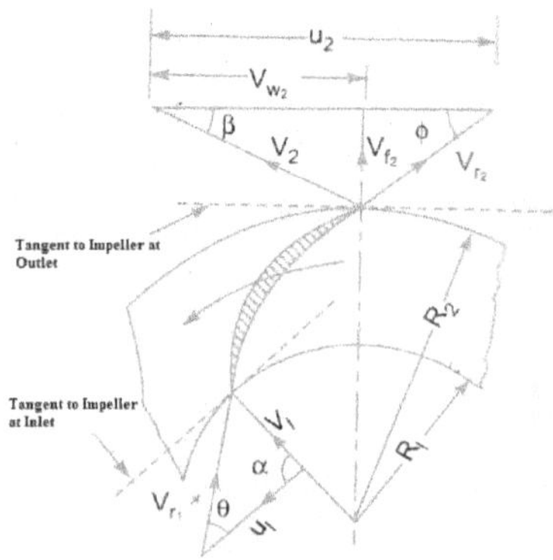

Let   N   = Speed of the impeller in rpm

D_1 = Diameter of impeller at inlet

u_1 = Tangential velocity of impeller at inlet

$$= \frac{\pi D_1 N}{60}$$

D_2 = Diameter of impeller at outlet

u_2 = Tangential velocity of impeller at outlet

$$= \frac{\pi D_2 N}{60}$$

V_1 = Absolute Velocity of water at inlet

V_{r_1} = Relative velocity of water at inlet.

$\propto$ = Angle made by absolute velocity (V_1) at inlet with the direction of motion of vane,

θ = Angle made by relative velocity (V_{r_1}) at inlet with the direction of motion of vane, and V_2, V_{r_2}, β and ϕ are the corresponding values at outlet.

As the water enters the impeller radially which means the absolute velocity of water at inlet is in the radial direction and hence angle $\propto = 90°$ and $V_{w_1} = 0$.

A centrifugal pump is the reverse of a radially inward flow reaction turbine. But in case of a radially inward flow reaction turbine, the work done by the water on the runner per second per unit weight of the water striking per second is given by the equation as

$$= \frac{1}{g}\left[V_{w_1} u_1 - V_{w_2} u_2\right]$$

Work done by the impeller on the water per second per unit weight of water striking per second

$$= \quad - [\text{Work done in case of turbine}]$$

$$=$$

$$-\left[\frac{1}{g}\left[V_{w_1}u_1 - V_{w_2}u_2\right]\right] = \frac{1}{g}\left[V_{w_2}u_2 - V_{w_1}u_1\right]$$

$$= \quad \frac{1}{g}V_{w_2}u_2$$

$$\left(\because V_{w_1} = 0 \ here\right) \qquad\qquad \text{(i)}$$

Work done by impeller on water per second.

$$= \quad \frac{W}{g}V_{w_2}u_2 \qquad\qquad \text{(ii)}$$

Where, W = Weight of water = $\rho \times g \times Q$

Q = Volume of water

And Q = Area x velocity of flow

$$= \quad \pi D_1 B_1 x V_{f_1} \quad = \quad \pi D_2 B_2 x V_{f_2} \quad \text{(iii)}$$

Where B_1 and B_2 are width of impeller at inlet and outlet and V_{f_1} and V_{f_2} are velocities of flow at inlet and outlet.

Equation (i) gives the head imparted to the water by the impeller or energy given by impeller to water per unit weight per second.

4.23. Efficiencies of a Centrifugal Pump

The power is transmitted from the shaft of the electric motor to the shaft of the pump and then to the impeller. From the impeller, the power is given to the water. Thus power is decreasing from the shaft of the pump to the impeller4 and them to the water. The followings are the important efficiencies of a centrifugal pump.

a) Manometric efficiency, η_{man}

b) Mechanical efficiency, η_m

c) Overall efficiency, η_o

a) Manometric efficiency (η_{man})

The ratio of the manometric head to the head imparted by the impeller to the water is known as manometric efficiency. Mathematically, it is written as

$$\eta_{man} = \frac{Manometric\ head}{Head\ imparted\ by\ impeller\ to\ water} = \frac{H_m}{\left(\dfrac{V_{w_2}u_2}{g}\right)} = \frac{gH_m}{V_{w_2}u_2}$$

The power at the impeller of the pump is more than the power given to the water at outlet of the pump. The ratio of the power given to water at outlet of the pump to the power available at the impeller, is known as manometric efficiency.

The power given to water at outlet of the pump $= \dfrac{WH_m}{1000}$ kW

$$\text{The power at the impeller} = \frac{Work\ done\ by\ impeller\ per\ second}{1000}\ \text{kW}$$

$$= \frac{W}{g} x \frac{V_{w_2}xu_2}{1000}\ \text{kW}$$

$$\eta_{man} = \frac{\dfrac{WxH_m}{1000}}{\dfrac{W}{g} x \dfrac{V_{w_2}xu_2}{1000}} = \frac{g\ x\ H_m}{V_{w_2}\ x\ u_2}$$

b) Mechanical Efficiency (η_m)

The power at the shaft of the centrifugal pump is more than the power available at the impeller of the pump. The ratio of the power available at the impeller to the power at the shaft of the centrifugal pump is known as Mechanical efficiency. It is written as

$$\eta_m = \frac{Power\ at\ the\ impeller}{Power\ at\ the\ shaft}$$

$$\text{The power at the impeller in kW} = \frac{Work\ done\ by\ impeller\ per\ second}{1000}$$

$$= \frac{W}{g} x \frac{V_{w_2}xu_2}{1000}$$

$$\eta_{man} = \frac{\dfrac{w}{g}\left(\dfrac{V_{w_2}u_2}{1000}\right)}{S.P}$$

Where S.P. $=$ Shaft power.

c) Overall Efficiency (η_o)

It is defined as ratio of power output of the pump to the power input to the pump.

The power output of the pump in kW. $= \dfrac{Weight\ of\ water\ lifted\ x\ H_m}{1000} = \dfrac{WH_m}{1000}$

Power input to the pump $=$ Power supplied by the electric motor.

$=$ S.P. of the pump

$$\eta_o = \dfrac{\left(\dfrac{WH_m}{1000}\right)}{S.P}$$

Also, $\eta_o = \eta_{man} \times \eta_m$

Problems in Centrifugal Pumps

1. The internal and external diameters of the impeller of a centrifugal pump are 200 mm and 400 mm respectively. The pump is running at 1200 rpm. The vane angles of the impeller at inlet and outlet are 20º and 30º respectively. The water enters the impeller radially and velocity of flow is constant. Determine the work done by the impeller per unit weight of water.

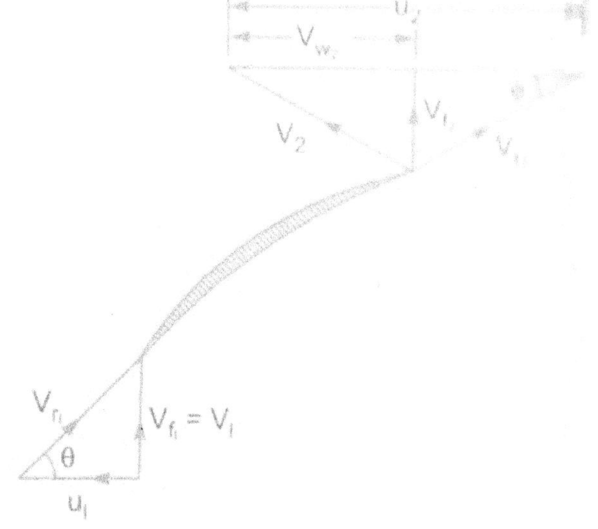

Given

Internal dia of impeller	D_1	=	200 mm =	0.2 m
External dia of impeller	D_2	=	400 mm =	0.4 m
Speed	N	=	1200 rpm	
Vane angle at inlet,	θ	=	20°	
Vane angle at outlet	ϕ	=	30°	
Water enters radially means $\propto$		=	90° and $V_{w_1} = 0$	
Velocity of flow	V_{f_1}	=	V_{f_2}	

Tangential velocity of impeller at inlet and outlet are,

$$u_1 = \frac{\pi D_1 N}{60} = \frac{\pi x 0.2 x 1200}{60} = 12.56 \text{ m/s}$$

and

$$u_2 = \frac{\pi D_2 N}{60} = \frac{\pi x 0.4 x 1200}{60} = 25.13 \text{ m/s}$$

From inlet velocity triangle, $\tan\theta = \dfrac{V_{f_1}}{u_1} = \dfrac{V_{f_1}}{12.56}$

$$V_{f_1} = 12.56 \tan\theta = 12.56 \times \tan 20$$

$$= 4.57 \text{ m/s}$$

$$V_{f_2} = V_{f_1} = 4.57 \text{ m/s}$$

From outlet velocity triangle, $\tan\theta = \dfrac{V_{f_2}}{u_2 - V_{w_2}} = \dfrac{4.57}{25.13 - V_{w_2}}$

OR

$$25.13 - V_{w_2} = \frac{4.57}{\tan\phi} = \frac{4.57}{\tan 30}$$

$$= 7.915$$

$$V_{w_2} = 25.13 - 7.915$$

$$= 17.215 \text{ m/s}$$

The work done by impeller per kg of water per second is given by equation as

$$= \frac{1}{g} V_{w_2} u_2 = \frac{17.215 x 25.13}{9.81}$$

$$= 44.1 \text{ Nm/N.} \quad \textbf{Ans}$$

2. A Centrifugal pump is to discharge 0.118 m³/s at a speed of 1450 rpm against a head of 25 m. The impeller diameter is 250 mm, its width at outlet is 50 mm and manometric efficiency is 75%. Determine the vane angle at the outer periphery of the impeller.

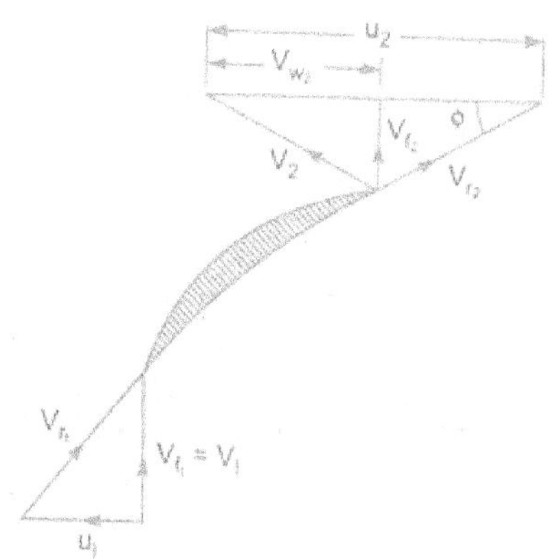

Given

Discharge	Q	=	0.118 m³/s
Speed	N	=	1450 rpm
Head	H_m	=	25 m
Diameter at outlet	D_2	=	250 mm = 0.25 m
Width at outlet	B_2	=	50 mm = 0.05 m
Manometric efficiency η_{man}		=	75% = 0.75
Let vane angle at outlet		=	ϕ

Tangential velocity of impeller at outlet,

$$u_2 = \frac{\pi D_2 N}{60} = \frac{\pi \times 0.25 \times 1450}{60}$$

$$= 18.98 \text{ m/s}$$

Discharge is given by

$$Q = \pi D_2 B_2 \times V_{f_2}$$

$$V_{f_2} = \frac{Q}{\pi D_2 B_2} = \frac{0.118}{\pi \times 0.25 \times 0.05}$$

$$= 3 \text{ m/s.}$$

Using equation

$$\eta_{man} = \frac{gH_m}{V_{w_2} u_2} = \frac{9.81 \times 25}{V_{w_2} \times 18.98}$$

$$V_{w_2} = \frac{9.81 x 25}{\eta_{man} x 18.98} = \frac{9.81 x 25}{0.75 x 18.98}$$

$$= 17.23$$

From outlet velocity triangle, we have

$$\tan \phi = \frac{V_{f_2}}{\left(u_2 - V_{w_2}\right)} = \frac{3.0}{(18.98 - 17.23)} = 1.7143$$

$$\phi = \tan^{-1} 1.7143 \quad = 59.74^o$$

$$\text{or } \mathbf{59^o\,44'} \qquad \mathbf{Ans.}$$

3. A centrifugal pump delivers water against a net head of 14.5 metres and a design speed of 1000 rpm. The vanes are curved back to an angle of 30° with the periphery. The impeller diameter is 300 mm and outlet width 50 mm. Determine the discharge of the pump if manometric efficiency is 95%.

Given

Net head $\qquad$ H_m $\qquad = \qquad$ 14.5 m

Speed $\qquad$ N $\qquad = \qquad$ 1000 rpm

Vane angle at outlet $\qquad$ ϕ $\qquad = \qquad$ 30°

Impeller diameter means the diameter of the impeller at outlet

$\qquad$ Diameter $\qquad$ D_2 $\qquad = \qquad$ 300 mm = 0.3 m

Outlet width $\qquad$ B_2 $\qquad = \qquad$ 50 mm = 0.05 m

Manometric efficiency η_{man} $\qquad = \qquad$ 95% = 0.95

Tangential velocity of impeller at outlet,

$$u_2 = \frac{\pi D_2 N}{60} = \frac{\pi x 0.3 x 1000}{60}$$

$$= 15.70 \text{ m/s}$$

Now using equation $\qquad \eta_{man} = \dfrac{gH_m}{V_{w_2} x u_2}$

$$0.95 = \frac{9.81 x 14.5}{V_{w_2} x 15.7}$$

$$V_{w_2} = \frac{0.95 x 14.5}{0.95 x 15.7}$$

$$= 9.54 \text{ m/s}$$

From outlet velocity triangle, we have

$$\tan \phi = \frac{V_{f_2}}{\left(u_2 - V_{w_2}\right)} \quad or \quad \tan 30^o = \frac{V_{f_2}}{(15.7 - 9.54)} = \frac{V_{f_2}}{6.16}$$

$$V_{f_2} = 6.16 \times \tan 30^o = 3.556 \text{ m/s}$$

$$\text{Discharge} \quad Q = \pi D_2 B_2 x V_{f_2} = \pi x 0.3 x 0.05 x 3.556$$

$$= \textbf{0.1675 m}^3\textbf{/s} \qquad \textbf{Ans.}$$

4. A centrifugal pump having outer diameter equal to two times the inner diameter and running at 1000 rpm works against a total head of 40 m. The velocity of flow through the impeller is constant and equal to 2.5 m/s. The vanes are set back at an angle of 40° at outlet. If the outer diameter of the impeller is 500 mm and width at outlet is 50 mm, Determine:

(i) Vane angle at inlet

(ii) Work done by impeller on water per second, and

(iii) Manometric efficiency.

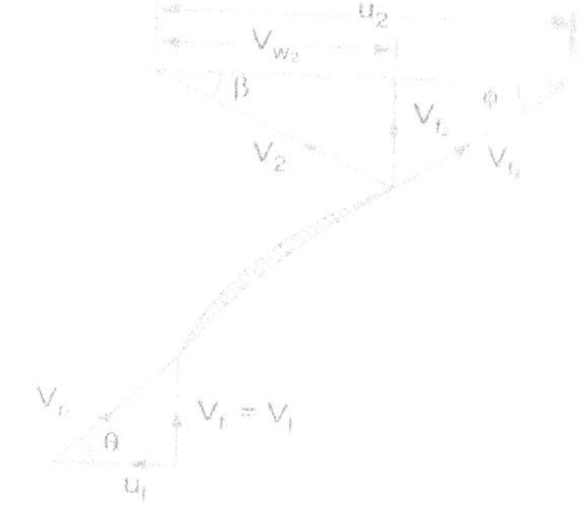

Given:

Speed   N   =   1000 rpm

Head H_m =   40 m

Velocity of flow V_{f_1} = V_{f_2} = 2.5 m/s

Vane angle at outlet ϕ =   40°

Outer dia of impeller D_2 =   500 mm = 0.5 m

Inner dia of impeller D_1 = $\dfrac{D_2}{2} = \dfrac{0.5}{2} = 0.25$ m

Width at outlet B_2 =   50 mm = 0.05 m

Tangential velocity of impeller at inlet and outlet are

$$u_1 \;=\; \frac{\pi D_1 N}{60} = \frac{\pi \times 0.25 \times 1000}{60}$$

$$= \; 13.09 \text{ m/s.}$$

$$u_2 \;=\; \frac{\pi D_2 N}{60} = \frac{\pi \times 0.5 \times 1000}{60}$$

$$= \; 26.18 \text{ m/s.}$$

Discharge is given by Q $= \; \pi D_2 B_2 x V_{f_2}$

$$= \; \pi \times 0.5 \times 0.05 \times 2.5$$

$$= \; 0.1963 \text{ m}^3/\text{s.}$$

(i) Vane angle at inlet (θ)

From inlet velocity triangle $\tan \theta \;=\; \dfrac{V_{f_1}}{u_1} = \dfrac{2.5}{13.09} \;=\; 0.191$

$$\theta \;=\; \tan^{-1} . 191 = 10.81^\circ$$

OR $10^\circ 48'$.   **Ans.**

(ii) Work done by impeller on water per second is given by equation as

$$= \; \frac{W}{g} x V_{w_2} u_2 = \frac{\rho x g x Q}{g} x V_{w_2} x u_2$$

$$= \; \frac{1000 x 9.81 x 0.1963}{9.81} x V_{w_2} x 26.18 \qquad \text{(i)}$$

But from outlet velocity triangle, we have

$$\tan \theta \;=\; \frac{V_{f_2}}{u_2 - V_{w_2}} = \frac{2.5}{\left(26.18 - V_{w_2}\right)}$$

$$26.18 - V_{w_2} \;=\; \frac{2.5}{\tan \phi} = \frac{2.5}{\tan\ 40} \;=\; 2.979$$

$$V_{w_2} \;=\; 26.18 - 2.979$$

$$= \; 23.2 \text{ m/s}$$

Substituting this value of V_{w_2} in equation (i), we get the work done by impeller as

$$= \; \frac{1000 x 9.81 x 0.1963}{9.81} x 23.2 x 26.18$$

$$= \; 119227.9 \text{ Nm/s} \;\; .... \text{ **Ans.**}$$

4.24. Derive the Expression for Specific Speed of the Turbine

Step 1

Definition

It is defined as the speed of a turbine which is identical in shape, geometrical dimensions, blade angles, gate opening etc., with the actual turbine but of such a size that it will develop unit power when working under unit heads. It is denoted by the symbol N_s. The specific speed is used in comparing the different types of turbines as every type of turbine has different specific speed.

The overall efficiency (η_o) of any turbine is given by

$$\eta_o = \frac{Shaft\ power}{Water\ Power} = \frac{Power\ developed}{\dfrac{\rho xgxQxH}{1000}} = \frac{P}{\dfrac{\rho xgxQxH}{1000}} \qquad (i)$$

$$
\begin{aligned}
\text{Where} \quad H &= \text{Head under which the turbine is working.}\\
Q &= \text{Discharge through turbine}\\
P &= \text{Power developed or shaft power.}
\end{aligned}
$$

$$\text{From equation (i),} \qquad P = \eta_o x \frac{\rho xgxQxH}{1000}$$

$$\propto x\ Q\ x\ H\ (as\ \eta_o\ and\ \rho\ are\ constant) \qquad (ii)$$

Step 2

$$
\begin{aligned}
\text{Now let} \quad D &= \text{Diameter of actual turbine.}\\
N &= \text{Speed of actual turbine}\\
u &= \text{Tangential velocity of the turbine,}\\
N_s &= \text{Specific speed of the turbine}\\
V &= \text{Absolute velocity of water.}
\end{aligned}
$$

Step 3

The absolute velocity, tangential velocity and head on the turbine are related as, $u \propto V$,

$$\text{where}\ V \propto \sqrt{H} \propto \sqrt{H} \qquad (iii)$$

Step 4

But the tangential velocity u is given by

$$U = \frac{\pi DN}{60} = \propto DN \qquad (iv)$$

Step 5

From equations (iii) and (iv) we have

$$\sqrt{H} \propto DN \quad \text{or} \quad D \propto \frac{\sqrt{H}}{N} \tag{v}$$

Step 6

The discharge through turbine is given by

$$Q = \text{Area} \times \text{Velocity} \quad (\text{where B = Width})$$

$$(B \propto D)$$

But

$$\text{Area} \propto B \times D$$

$$\propto D^2$$

And

$$\text{Velocity} \propto \sqrt{H}$$

$$Q \propto D^2 \times \sqrt{H}$$

$$\propto \left(\frac{\sqrt{H}}{N}\right)^2 \times \sqrt{H} \qquad \left(\because \text{From equation}(v), D \propto \frac{\sqrt{H}}{N}\right)$$

$$\propto \frac{H}{N^2} \times \sqrt{H} \propto \frac{H^{3/2}}{N^2} \tag{vi}$$

Step 7

Substituting the value of Q in equation (ii), we get

$$P \propto \frac{H^{3/2}}{N^2} \times H \propto \frac{H^{5/2}}{N^2}$$

$$P = K \frac{H^{5/2}}{N^2}, \text{ where K = Constant of proportionality}$$

If P = 1, H = 1, the speed N = Specific speed N_s. Substituting these values in the above equation, we get,

$$1 = K \frac{1^{5/2}}{N_s^2} \qquad \text{or} \qquad N_s^2 = K$$

$$P = N_s^2 \frac{H^{5/2}}{N^2} \qquad \text{or} \qquad N_s^2 = \frac{N^2 P}{H^{5/2}}$$

$$N_s = \sqrt{\frac{N^2 P}{H^{5/2}}} = \frac{N\sqrt{P}}{H^{5/4}}$$

If P is taken in metric horse power the specific speed is obtained in M.K.S. Units. But if P is taken in kilowatts, the specific speed is obtained in S.I. units.

4.25. Significance of Specific Speed

Specific speed plays an important role for selecting the type of the turbine. Also the performance of a turbine can be predicted by knowing the specific speed of the turbine. The type of turbine for different specific speed is given in table as

Table: Range of Specific speed of the turbine

S. No.	Specific Speed		Types of turbine
	M.K.S.	S.I.	
1	10 to 35	8.5 to 30	Pelton wheel with single jet
2	35 to 60	30 to 51	Pelton wheel with two or more jets
3	60 to 300	51 to 225	Francis Turbine
4	300 to 1000	255 to 860	Kaplan or Propeller turbine

4.26. Unit Quantities of Turbine and Pump

Sl. No.	Description	Turbine	Pump
1.	Specific speed	$N_s = \dfrac{N\sqrt{P}}{H^{5/4}}$	$N_s = \dfrac{N\sqrt{Q}}{H_m^{3/4}}$
2.	Unit speed	$\dfrac{N_1}{\sqrt{H_1}} = \dfrac{N_2}{\sqrt{H_2}}$	$\dfrac{\sqrt{H_{m_1}}}{D_1 N_1} = \dfrac{\sqrt{H_{m_2}}}{D_2 N_2}$
3.	Unit Discharge	$\dfrac{Q_1}{\sqrt{H_1}} = \dfrac{Q_2}{\sqrt{H_2}}$	$\dfrac{Q_1}{D_1^{3} N_1} = \dfrac{Q_2}{D_2^{3} N_2}$
4.	Unit power	$\dfrac{P_1}{H_1^{3/2}} = \dfrac{P_2}{H_2^{3/2}}$	$\dfrac{P_1}{D^5 N_1^{3}} = \dfrac{P_2}{D_2^{5} N_2^{3}}$

1. A turbine develops 9000 kW when running at a speed of 140 rpm and under a head of 30 m. Determine the specific speed of the turbine.

Given

Power developed	P	=	9000 kW
Head	H	=	30 m
Speed,	N	=	140 rpm

The specific speed is given by equation as

$$N_s = \frac{N\sqrt{P}}{H^{5/4}} = \frac{140 x \sqrt{9000}}{30^{5/4}} = \frac{13281.56}{70.21}$$

$$= 189.167 \text{ SI Units} \qquad \textbf{Ans.}$$

Unit Quantities

In order to predict the behavior of a turbine working under varying conditions of head, speed, output and gate opening, the results are expressed in terms of quantities which may be obtaine3d when the head on the turbine is reduced to unity. The conditions of the turbine under unit head are such that the efficiency of the turbine remains unaffected. The followings are the three important unit quantities which must be studied under unit head:

1. Unit Speed.
2. Unit Power.
3. Unit discharge.

1. Unit Speed

It is defined as the speed of a turbine working under a unit head (i.e. under a head of 1 m). It is denoted by 'N_u'. The expression for unit speed (N_u) is obtained as:

$$\text{Let,} \quad N \quad = \quad \text{Speed of a turbine under a head H}$$
$$H \quad = \quad \text{Head under which a turbine is working.}$$
$$u \quad = \quad \text{Tangential Velocity.}$$

The tangential velocity, absolute velocity of water and head on the turbine are related as

$$u \quad \propto \quad V \qquad \text{where } V \propto \sqrt{H}$$
$$\propto \quad \sqrt{H} \qquad \qquad \text{(i)}$$

Also tangential velocity (u) is given by

$$u \quad = \quad \frac{\pi DN}{60} \qquad \text{where } D = \text{Diameter of turbine.}$$

For a given turbine, the diameter (D) is constant.

$$u \propto N \quad \text{or} \quad N \propto u \quad \text{or} \quad N \propto \sqrt{H}$$

$$(\because From(i), u \infty \sqrt{H})$$

$$N \quad = \quad K_1 \sqrt{H} \qquad \text{(ii)}$$

Where K_1 is a constant of proportionality.

If head on the turbine becomes unity, the speed becomes unit speed or

$$\text{When} \quad H \quad = \quad 1, \quad N \quad = \quad N_u$$

Substituting these values in equation (ii), we get

$$N_u \quad = \quad K_1 \sqrt{1.0} \quad = \quad K_1$$

Substituting the value of K_1 in equation (ii)

$$N = N_u \sqrt{H} \quad \text{or} \quad N_u = \frac{N}{\sqrt{H}}$$

2. Unit Power

It is defined as the power developed by a turbine, working under a unit head (i.e. under a head or 1 m). It is denoted by the symbol P_u. The expression for unit power is obtained as:

Let, H = Head of water on the turbine

P = Power developed by the turbine under a head of H

Q = Discharge through turbine under a head H

The overall efficiency (η_o) is given as

$$\eta_o = \frac{Power\ developed}{Water\ power} = \frac{P}{\dfrac{\rho x g x Q x H}{1000}}$$

$$P = \eta_o x \frac{\rho x g x Q x H}{1000}$$

$$\propto Q \times H$$

$$\propto \sqrt{H x H} \qquad\qquad \left(\because Q \infty \sqrt{H} \right)$$

$$\propto H^{3/2}$$

$$P = K_3 H^{3/2} \qquad\qquad \text{(iv)}$$

Where K_3 is a constant of proportionality,

When H = $1m$ P = P_u

$$P_u = K_3 (1)^{3/2} = K_3$$

Substituting the value of K_3 in equation (iv), we get

$$P = P_u H^{3/2}$$

$$P_u = \frac{P}{H^{3/2}}$$

3. Unit Discharge

It is defined as the discharge passing through a turbine, which is working under a unit head (i.e., 1 m). It is denoted by the symbol 'Q_u'. The expression for unit discharge is given as

Let, H = Head of water on the turbine,

Q = Discharge passing through turbine when head is H on the turbine

$\propto$ = Area of flow of water.

The discharge passing through a given turbine under a head 'H' is given by,

$$Q \quad = \quad \text{Area of flow x velocity.}$$

But for a turbine, area of flow is constant and velocity is proportional to $\sqrt{H}$

$$Q \propto \text{Velocity} \propto \sqrt{H}$$

OR $\qquad\qquad Q \quad = \quad K_{2 \, x} \, H \qquad\qquad\qquad\qquad$ (iii)

Where K_2 is constant of proportionality,

If $\qquad\qquad H \quad = \quad 1. \qquad Q = Q_u \qquad$ (By definition)

Substituting these values in equation (iii), we get

$$Q_u \quad = \quad K_2 \sqrt{1.0} \quad = \quad K_2$$

Substituting the value of K_2 in equation (iii) we get

$$Q \quad = \quad Q_u \sqrt{H}$$

$$Q_u \quad = \quad \frac{Q}{\sqrt{H}}$$

2. A turbine develops 9000 kW when running at 10 rpm. The head on the turbine is 30 m. If the head on the turbine is reduced to 18 m., determine the speed and power developed by the turbine.

Given

Power developed	P_1	=	9000 kW
Speed	N_1	=	100 rpm
Head	H_1	=	30 m
Let for a head,	H_2	=	18 m
Speed	N_2		
Power	P_2		

Using equation, $\quad \dfrac{N_1}{\sqrt{H_1}} = \dfrac{N_2}{\sqrt{H_2}}$

$$N_2 \quad = \quad \frac{N_1 \sqrt{H_2}}{\sqrt{H_1}} = \frac{100\sqrt{18}}{\sqrt{30}} = \frac{100 x 4.2426}{5.4772}$$

$$= \quad 77.46 \quad \text{rpm} \quad \textbf{Ans.}$$

Also we have
$$\frac{P_1}{H_1^{3/2}} = \frac{P_2}{H_2^{3/2}}$$

$$P_2 = \frac{P_1 H_2^{3/2}}{H_1^{3/2}} = \frac{9000 x 18^{3/2}}{30^{3/2}} = \frac{687307.78}{164.316}$$

$$= \quad 4182.84 \text{ kW} \quad \textbf{Ans.}$$

3. A turbine develops 500 kW power under a head of 100 metres at 200 rpm. What would be its normal speed and output under a head of 81 metres?

Given

Power	P_1	=	500 kW
Head	H_1	=	100 m
Speed	N_1	=	200 rpm
For a head,	H_2	=	81 m
Let,	N_2	=	Speed
	P_2	=	Power.

Using equations for speed, we have

$$\frac{N_1}{\sqrt{H_1}} = \frac{N_2}{\sqrt{H_2}}$$

$$N_2 = \sqrt{H_2}\frac{N_1}{\sqrt{H_1}} = \sqrt{\frac{H_2}{H_1}} x N_1 = \sqrt{\frac{81}{100}} x 200$$

$$= \frac{9}{10} x 20$$

$$= \quad 180 \quad \text{rpm} \quad \textbf{Ans.}$$

Using equation for power, we have

$$\frac{P_2}{H_1^{3/2}} = \frac{P_2}{H_2^{3/2}}$$

$$P_2 = H_2^{3/2} x \frac{P_2}{H_1^{3/2}} = \frac{81^{3/2}}{100^{3/2}} x 500$$

$$= \frac{729}{1000} x 500$$

$$= \quad 364.5 \text{ kW} \quad \textbf{Ans.}$$

4. A turbine is to operate under a head of 25 m at 200 rpm. The discharge is 9 cumec. If the efficiency is 90%, determine the performance of the turbine under a head of 20 metres.

Given

Head on turbine, H_1 = 25 m

Speed N_1 = 200 rpm

Discharge Q_1 = 9 m³/s

Overall efficiency η_o = 90% or 0.9

Performance of the turbine under a head, H_2 = 20 m, means to find the speed, discharge and power developed by the turbine when working under the head of 20 m.

Let for the head, H_2 = 20 m,

Speed = N_2, Discharge = Q_2 and Power = P_2

Using the relation, η_o = $\dfrac{P}{W.P.} = \dfrac{P_1}{\dfrac{\rho x g x Q_1 x H_1}{1000}}$

$$P_1 = \frac{\eta_o x \rho x g x Q_1 x H_1}{1000} = \frac{0.9x1000x9.81x9x25}{1000}$$

$$= 1986.5 \text{ kW}$$

Using equation $\dfrac{N_1}{\sqrt{H_1}} = \dfrac{N_2}{\sqrt{H_2}}$

$$N_2 = \frac{N_1\sqrt{H_2}}{\sqrt{H_1}} = 200x\frac{\sqrt{20}}{\sqrt{25}}$$

$$= 178.88 \text{ rpm} \qquad \textbf{Ans.}$$

Also, $\dfrac{Q_1}{\sqrt{H_1}} = \dfrac{Q_2}{\sqrt{H_2}}$

$$Q_2 = \frac{\sqrt{Q_1}}{\sqrt{H_1}} = 9.0x\sqrt{\frac{20}{25}}$$

$$= 8.05 \quad \text{m³/s.} \qquad \textbf{Ans.}$$

And
$$\frac{P_1}{H_1^{3/2}} = \frac{P_2}{H_2^{3/2}}$$

$$P_2 = \frac{P_1 H_2^{3/2}}{H_1^{3/2}} = P_1\left(\frac{H_2}{H_1}\right)^{3/2} = 1986.5\left(\frac{20}{25}\right)^{3/2}$$

$$= 1421.42 \text{ kW. } \textbf{Ans.}$$

5. A Pelton wheel is revolving at a speed of 190 rpm and develops 5150.25 kW when working under a head of 220 m with an overall efficiency of 80%. Determine unit speed, unit discharge and unit power. The speed ratio for the turbine is given as 0.47. Find the speed, discharge and power when this turbine is working under a head of 140 m.

Given

Speed	N_1	=	190 rpm
Power	P_1	=	5150.25 kW
Head	H_1	=	220 m
Overall efficiency	η_o	=	80% = 0.8
Speed ratio		=	0.47
New head of water	H_2	=	140 m

Overall efficiency is given by $\eta_o = \dfrac{P_1}{\dfrac{\rho x g x Q_1 x H_1}{1000}} = \dfrac{1000 x P_1}{\rho x g x Q_1 x H_1}$

$$Q_1 = \frac{1000 x P_1}{\eta_o x \rho x g x H_1} = \frac{1000 x 5150.25}{0.8 x 1000 x 9.81 x 220}$$

$$= 2.983 \text{ m}^3/\text{s}$$

Unit speed is given by equation, $\quad N_u \quad = \quad \dfrac{N_1}{\sqrt{H_1}} = \dfrac{190}{\sqrt{220}}$

$$= 12.81 \text{ rpm} \qquad \textbf{Ans.}$$

Unit discharge is given by equation,

$$Q_u = \frac{Q_1}{\sqrt{H_1}} = \frac{2.983}{\sqrt{220}} = 0.201 \text{ m}^3/\text{s} \qquad \textbf{Ans.}$$

Unit power is given by equation,

$$P_u = \frac{P_1}{H_1^{3/2}} = \frac{5150.25}{220^{3/2}} = 1.578 \text{ kW} \qquad \textbf{Ans.}$$

When the turbine is working under a new head of 140 m, the speed, discharge and power are given by equation

For speed

$$\frac{N_1}{\sqrt{H_1}} = \frac{N_2}{\sqrt{H_2}}$$

$$N_2 = \frac{N_1\sqrt{H_2}}{\sqrt{H_1}} = N_1\sqrt{\frac{H_2}{H_1}} = 190\sqrt{\frac{140}{220}}$$

$$= 151.56 \text{ rpm} \qquad \textbf{Ans.}$$

For discharge,

$$\frac{Q_1}{\sqrt{H_1}} = \frac{Q_2}{\sqrt{H_2}}$$

$$Q_2 = \frac{Q_1\sqrt{H_2}}{\sqrt{H_1}} = Q_1\sqrt{\frac{H_2}{H_1}} = 2.983\sqrt{\frac{140}{220}}$$

$$= 2.379 \text{ m}^3/\text{s} \qquad \textbf{Ans.}$$

For power,

$$\frac{P_1}{H_1^{3/2}} = \frac{P_2}{H_2^{3/2}}$$

$$P_2 = P_1\frac{H_2^{3/2}}{H_1^{3/2}} = P_1\left(\frac{H_2}{H_1}\right)^{3/2} = 5150.25\left(\frac{140}{220}\right)^{3/2}$$

$$= 2614.48 \text{ kW} \qquad \textbf{Ans.}$$

6. The diameter of a centrifugal pump, which is discharging 0.03 m³/s of water against a total head of 20 m is 0.4 m. The pump is running at 1500 rpm. Find the head, discharge and ratio of powers of a geometrically similar pump of diameter 0.25 m when it is running at 3000 rpm.

Given

Centrifugal pump

Discharge	Q_1	=	0.03 m³/s
Head	H_{m_1}	=	20 m
Diameter	D_1	=	0.4 m
Speed,	N_1	=	1500 rpm

Geometrically similar pump,

Diameter,	D_2	=	0.25 m
Speed	N_2	=	3000 rpm

Let Head on similar group $\qquad = \qquad H_{m_2}$

Discharge on similar pump $\qquad = \qquad Q_2$

Using equation, $\left(\dfrac{Q}{D^3 N}\right)_1 = \left(\dfrac{Q}{D^3 N}\right)_2$

$$\frac{Q_1}{Q_1^{\,3} N_1} = \frac{Q_1}{D_2^{\,3} N_2}$$

$$\frac{0.03}{0.4^3 \, x1500} = \frac{Q_2}{0.25^3 \, x3000}$$

$$Q_2 = \frac{0.3 x 0.25^2 \, x3000}{0.4^3 \, x1500} = 0.03x\left(\frac{0.25}{0.4}\right)^3 x2.0$$

$$= 0.01465 \text{ m}^3/\text{s} \qquad \textbf{Ans.}$$

Using equation, we have

$$\left(\frac{\sqrt{H_m}}{DN}\right)_1 = \left(\frac{\sqrt{H_m}}{DN}\right)_2$$

OR $\dfrac{\sqrt{H_{m_1}}}{D_1 N_1} = \dfrac{\sqrt{H_{m_2}}}{D_2 \, N_2} \qquad \therefore \dfrac{\sqrt{20}}{0.4 x1500} x \dfrac{\sqrt{H_{m_2}}}{0.25 x3000}$

OR $\sqrt{H_{m_2}} = \dfrac{\sqrt{20} x 0.25 x 3000}{0.4 x 1500} \qquad = \qquad 5.59$

$H_{m_2} = \qquad (5.59)^2$

$\qquad = \qquad 31.25$ m **Ans.**

Using equation, we have

$$\left(\frac{P}{D^5 N^3}\right)_1 = \left(\frac{P}{D^5 N^3}\right)_2$$

$$\frac{P}{D_1^{\,5} N_1^{\,3}} = \frac{P_2}{D_2^{\,5} N_2^{\,3}} \; or \; \frac{P_1}{P_2} = \frac{D_1^{\,5} N_1^{\,3}}{D_2^{\,5} N_2^{\,3}} = \left(\frac{D_1}{D_2}\right)^5 x\left(\frac{N_1}{N_2}\right)^3$$

$$= \left(\frac{0.4}{0.25}\right)^5 x\left(\frac{1500}{3000}\right)^3$$

$$= 10.485 \text{ x } 0.125 = 1.31 \qquad \textbf{Ans.}$$

4.27. Characteristic Curves of Hydraulic Turbines

Characteristic curves of a hydraulic turbine are the curves, with the help of which the exact behavior and performance of the turbine under different working conditions, can be known. These curves are plotted from the results of the tests performed on the turbine under different working conditions.

The important parameters which are varied during a test on a turbine are –

1. Speed (N)
2. Head (H)
3. Discharge (Q)
4. Power (P)
5. Overall efficiency (η_o)
6. Gate opening

Out of the above six parameters, three parameters namely speed (N), Head (H) and discharge (Q) are independent parameters.

Out of the three independent parameters, (N, H, Q) one of the parameter is kept constant (say H) and the variation of the other four parameters with respect to any one of the remaining two independent variables (say N and Q) are plotted and various curves are obtained. These curves are called characteristic curves. The followings are the important characteristic curves of a turbine.

1. Main characteristic curves or Constant Head Curve.
2. Operating Characteristic Curves or Constant Speed Curve.
3. Muschel Curves or Constant Efficiency Curve.

1. Main characteristic curves or Constant Head Curves

Main characteristic curves are obtained by maintaining a constant head and a constant gate opening (G.O.) on the turbine. The speed of the turbine is varied by changing load on the turbine. For each value of the speed, the corresponding values of the power (P) and discharge (Q) are obtained. Then the overall efficiency (η_o) for each value of the speed is calculated. From these readings the values of unit speed (N_u) unit power (P_u) and unit discharge (Q_u) are determined. Takin g N_u) as abscissa, the values of Q_u, P_u, P and η_o are plotted as shown in figure. By changing the gate opening, the values of Q_u, P_u and η_o and N_u are determined and taking N_u as abscissa, the values of Q_u, P_u and η_o are plotted. Figure shows the main characteristic curves for Pelton wheel and Figure shows the main characteristic curves for reaction (Francis and Kaplan) turbine.

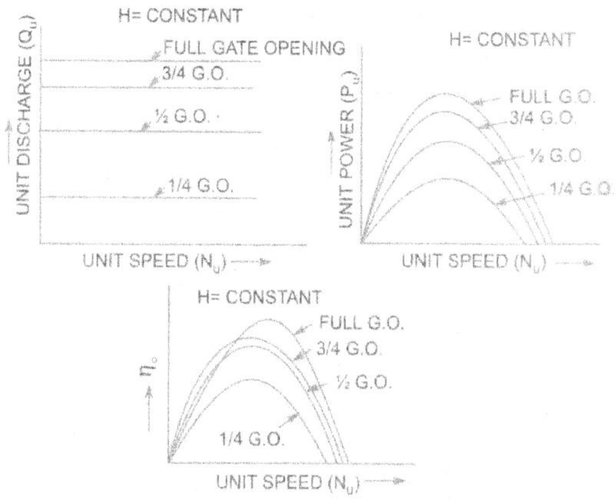

Main Characteristic Curves for a Pelton Wheel

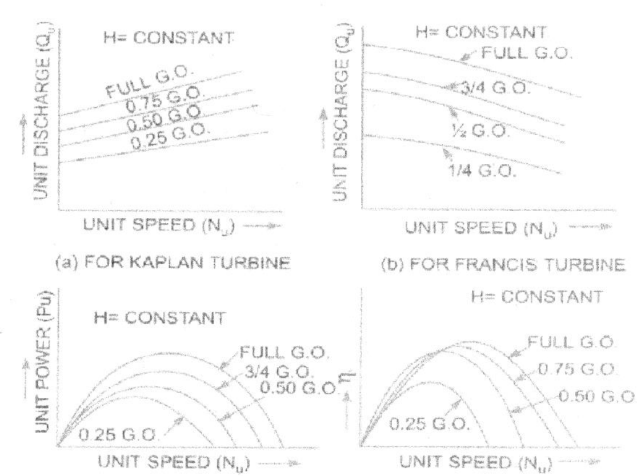

Main Characteristic Curves for Reaction Turbine

2. Operating Characteristic Curves or Constant Speed Curves

Operating Characteristic curves are plotted when the speed on the turbine is constant. In case of turbines, the head is generally constant. There are three independent parameters namely N, H and Q. For operating characteristics N and H are constant and hence the variation of power and efficiency with respect to discharge Q are plotted. The power curve for turbines shall not pass through the origin because certain amount of discharge is needed to produce power to overcome initial friction. Hence, the power and efficiency curves will be slightly away from the origin on the x-axis, as to overcome initial friction certain amount of discharge will be required. Figure shows the variation of power and efficiency with respect to discharge.

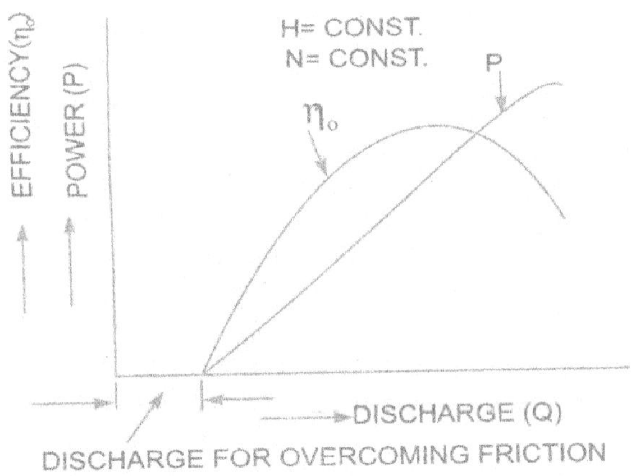

Operating Characteristic Curves

3. Constant Efficiency Curves or Muschel Curves or Iso-Efficiency Curves

These curves are obtained from the speed vs. efficiency and speed vs. discharge curves for different gate openings. For a given efficiency from the N_u vs. η_o curves, there are two speeds. From the N_u vs. Q_u curves, corresponding to two values of speeds there are two values of discharge. Hence, for a give efficiency there are two values of discharge for a particular gate opening. This means for a given efficiency there are two values of speeds and two values of the discharge for a given gate opening. If the efficiency is maximum there is only one value. These two values of speed and two values of discharge corresponding to a particular gate opening are plotted as shown in figure. The procedure is repeated for different gate openings and the curves Q vs N are plotted. The points having the same efficiencies are joined. The curves having same efficiency are called iso-efficiency curves. These curves are helpful for determining the zone of constant efficiency and for predicating the performance of the turbine at various efficiencies.

For plotting the iso-efficiency curves, horizontal lines representing the same efficiency are drawn on the η_o~speed curves. The points, at which these lines cut the efficiency curves at various gate openings are transferred to the corresponding Q~speed curves. The points having the same efficiency are then joined by a smooth curves. These smooth curves represents the iso-efficiency curve.

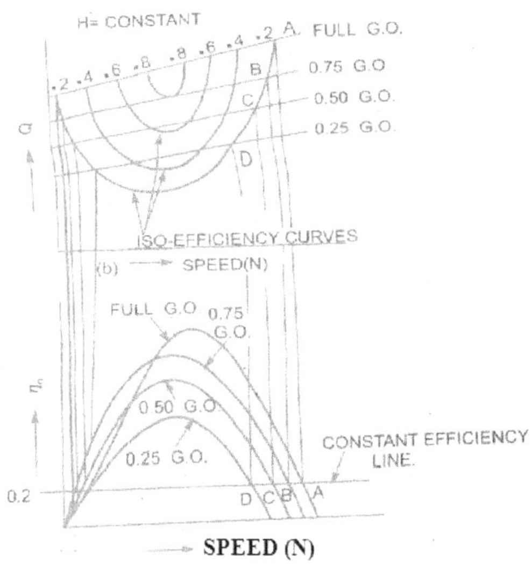

Constant efficiency Curve

Governing of Turbines

The governing of a turbine is defined as the operation by which the speed of the turbine is kept constant under all conditions of working. It is done automatically by means of a governor, which regulates the rate of flow through the turbines according to the changing load conditions on the turbine.

Governing of a turbine is necessary as a turbine is directly coupled to an electric generator, which is required to run at constant speed under all fluctuating loads conditions. The frequency of power generation by a generator of constant number of pair of poles under all varying conditions should be constant. This is only possible when the speed of the generator, under all changing load condition, is constant. The speed of the generator will be constant, when the speed of the turbine (which is coupled to the generator) is constant.

When the load on the generator decreases the speed of the generator increases beyond the normal speed (constant speed). When the speed of the turbine also increases beyond the normal speed. If the turbine or the generator is to run at constant (normal) speed, the rate of flow of water to the turbine should be decreased it-till the speed becomes normal. This process by which the speed of the turbine (and hence of generator) is kept constant under varying condition of load is called governing.

Governing of Pelton Turbine (Impulse Turbine)

Governing of Pelton turbine is done by means of oil pressure governor, which consists of the following parts:

1. Oil sump
2. Gear pump also called oil pump, which is driven by the power obtained from turbine shaft.
3. The Servomotor also called the relay cylinder.
4. The control valve or the distribution valve or relay valve.
5. The centrifugal governor or pendulum which is driven by belt or gear from the turbine shaft.
6. Pipes connecting the oil sump with the control valve and control valve with servomotor and
7. The spear rod or needle.

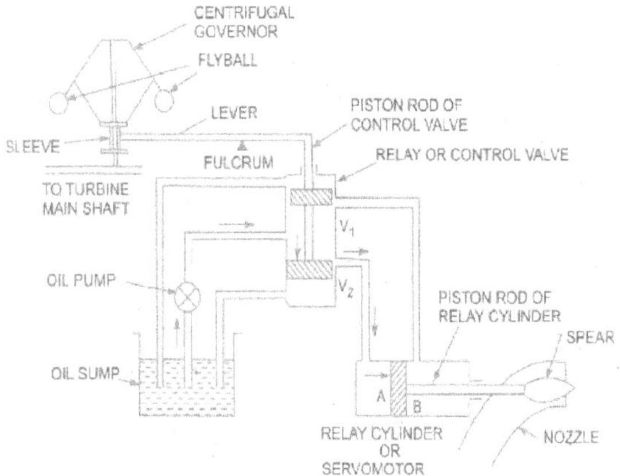

Governing of Pelton Turbine

Figure shows the position of the piston in the relay cylinder, position of control or relay valve and fly-balls of the centrifugal governor, when the turbine is running at the normal speed. When the load on the generator decreases, the speed of the generator increases. This increases the speed of the turbine beyond the normal speed. The centrifugal governor, which is connected to the turbine main shaft, will be rotating at an increased speed. Due to increase in the speed of the centrifugal governor, the fly-balls move upward due to the increased centrifugal force on them. Due to the upward movement of the fly-balls, the sleeve will also move upward. A horizontal lever, supported over a fulcrum, connects the sleeve and the piston rod of the control valve. As the sleeve moves up, the lever turns about the fulcrum and the piston rod of the control valve moves downward. This closes the valve V_1 and opens the valve V_2 as shown in figure.

The oil, pumped from the oil pump to the control valve or relay valve, under pressure will flow through the valve V_2 to the servomotor (or relay cylinder) and will exert force on the face A of the piston of the relay cylinder. The piston along with piston rod and spear will move towards right. This will decrease the area of flow of water at the outlet of the nozzle. This decrease of area of flow will reduce the rate of flow of water to the turbine which consequently reduces the speed of the turbine. When the speed of the turbine becomes normal, the fly-balls, sleeve, lever and piston rod of control valve come to its normal position as shown in figure.

When the load on the generator increases, the speed of the generator and hence of the turbine decreases.

The speed of the centrifugal governor also decreases and hence centrifugal force acting on the fly-balls also reduces.

This brings the fly-balls in the downward direction. Due to this, the sleeve moves downward and the lever turns about the fulcrum, moving the piston rod of the control valve in the upward direction. This closes the valve V_2 and opens the valve V_1.

The oil under pressure from the control valve, will move through valve V_1 to the servomotor and will exert a force on the face B of the piston. This will move the piston along with the piston rod and spear towards left, increasing the area of flow of water at the outlet of the nozzle.

This will increase the rate of flow of water to the turbine and consequently, the speed of the turbine will also increase, till the speed of the turbine becomes normal.

Two Marks Question-Pumps and Turbines

Hydraulic Turbines

1. Write the Euler's equation for Turbo machines?

   Momentum of the water $\qquad = \rho a V_1(V_{W_1} + V_{W_2})xu$

   ρ - Density; a – area of jet; V1 – Jet velocity; V_{W1} and V_{W2} – Whirl velocity Inlet and Outlet; u – Bucket speed

2. What are the components of energy transfer in turbo machines?

   1. Whirl velocity.

   2. Flow velocity.

   3. Relative velocity.

   Define degree of reaction.

   $$\text{Degree of Reaction (R)} = \frac{\textit{Increase in relative kinetic energy in moving blade}}{\textit{Stage Work output}}$$

3. Give an example for a low head turbine, a medium head turbine and a high head turbine.

High head turbine	Medium head turbine	Low head turbine
Head above 250 m	Head 60 m to 250 m	Head less than 60 m
Ex. Pelton wheel	Ex. Francis turbine	Ex. : Kaplan turbine.

4. Differentiate between impulse and reaction turbine.

Impulse Turbine	Reaction Turbine
If at the inlet of the turbine, the energy available is only kinetic energy the turbine is known as Impulse turbine	If at the inlet of the turbine, the water possesses kinetic energy as well as pressure energy the turbine is known as Reaction turbine
Blades are only in action when they are in front of nozzle.	Blades are in action at all the time.
Example: Pelton wheel turbine	Example: Francis turbine, Kaplan turbine

5. Define hydraulic, Mechanical, Volumetric and Overall efficiency.

   Hydraulic efficiency (η_h) $\quad \dfrac{R.P.}{W.P.}$ (R.P.- Runner power; W.P.- Water power)

   Mechanical efficiency (η_h) $\quad \dfrac{S.P.}{R.P.}$ (S.P.- Shaft power)

   Volumetric efficiency (η_v) $\quad = \dfrac{\textit{Volume of water actually striking the runner}}{\textit{Volume of water sup plied to the turbine}}$

Overall efficiency $(\eta_o)= \dfrac{S.P.}{W.P.} = \dfrac{P}{\dfrac{\rho g Q H}{1000}} x100$

6. What is draft tube?

    The tube which increases the outlet velocity of turbines is known as draft tube. So, head is saved by fitting draft tube.

7. Write the function of draft tube in turbine outlet?

    1. It allows the turbine to be sent above tail – water level without loss of head for doing inspection and maintenance.
    2. It regains the major portion of kinetic energy delivered from the runner by diffuse action.

8. Define specific speed of a turbine.

    It is defined as the speed of a geometrically similar turbine (i.e. a turbine identical in shape, dimensions, blade angles etc.,) which will develop unit power when working under unit head.

    $$Ns = \text{Specific speed of Turbine} = \frac{N\sqrt{P}}{H^{\frac{5}{4}}}$$

9. Define Unit Speed of turbine; Unit Discharge of turbine and Unit power of turbine.

    Unit Speed of turbine: $\qquad N_u \quad = \quad \dfrac{N}{\sqrt{H}.}$

    Unit Discharge of turbine: $\qquad Q_u \quad = \quad \dfrac{Q}{\sqrt{H}.}$

    Unit power of turbine $\qquad P_u \quad = \quad \dfrac{P}{(H)^{3/8}}$

10. Give the relationship between the jet speed and bucket speed.

    $u = \dfrac{V_1}{2.}$ u – Bucket speed; V_1 - Jet speed.

11. Difference between pump and turbine.

Turbine	Pump
The turbine converts hydraulic energy into mechanical energy. This mechanical energy is converted into electrical energy. Example: Pelton Wheel, Kaplan turbine; Francis Turbine.	A pump is a device which converts mechanical energy into hydraulic energy. Example: Centrifugal pump; Reciprocating pump.

12. Classify turbines according to flow (or) classification of hydraulic turbine.

1. According to the water flowing / energy inlet (Impulse turbine; Reaction turbine)

2. According to the direction of flow through runner.

   (i) Tangential flow turbine.

   (ii) Radial flow turbine.

   (iii) Axial flow turbine.

   (iv) Mixed flow turbine.

3. According to the head at the inlet of turbine

   a) High head turbine.

   b) Medium head turbine.

   c) Low head turbine.

4. According to the specific speed of the turbine.

   a) Low specific speed of turbine.

   b) Medium specific speed of turbine.

   c) High specific speed of turbine.

13. What is NPSH?

> **NPSH = Net Positive Suction Head**

NPSH = Absolute Pressure head @ inlet of the pump – Vapour pressure head +velocity head

Hydraulic Pumps

1. List out various components of a Centrifugal pump.

   a. Impeller.

   b. Casing.

   c. Suction pipe, Foot valve.

   d. Delivery pipe and Delivery valve.

2. What are the various types of casing?

   a. Volute Casing.

   b. Vertex Casing.

   c. Volute casing with guide blade.

3. What is meant by priming?

The delivery valve is closed and the suction pipe, casing and portion of the delivery pipe upto delivery valve are completely filled with the liquid, so that no air pocket is left. This is called as priming. (Removing the airlock)

4. Write down the formula for manometric efficiency.

$$\eta_{mano} = \frac{Manometric\ head}{Head\ imparted\ by\ impeller\ to\ liquid}$$

5. Define speed ratio and flow ratio.

Speed Ratio	Flow Ratio
$u = \phi\sqrt{2gH}$	$V_{f_1} = (K_f)x\sqrt{2gH}$
ϕ = Speed Ratio	K_f = Flow ratio

6. Mention the main components of reciprocating pump.

   a) Piston or Plunger.

   b) Suction valve.

   c) Delivery valve.

   d) Crank and connecting rod.

7. Define "Slip" of Reciprocating pump. When does the negative slip occur?

   Slip = Theoretical Discharge (Q_{th}) – Actual Discharge (Q_{act}).

$$Cd = \frac{Q_{act}}{Q_{theo}}$$

But in some times Q_{act} may be higher than Q_{th}, in such case Cd is greater than unity and slip will be negative called as negative slip.

8. Define indicator diagram.

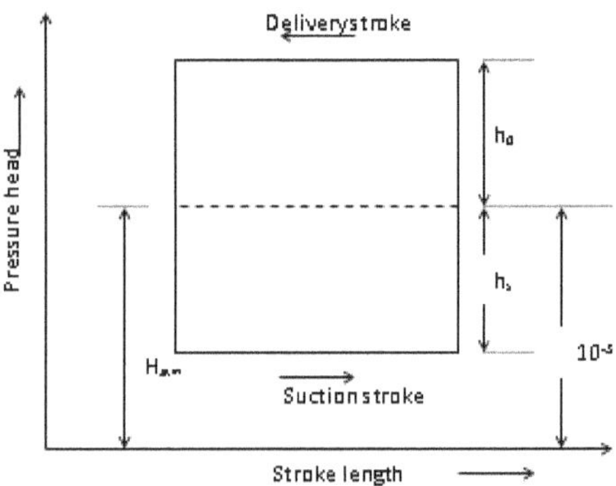

Indicator diagram is nothing but a graph plotted between the pressure head in the cylinder and the distance traveled by piston from Inner dead centre for one complete revolution of the crank.

9. Write down the formula for specific speed of a pump

$$N_s \quad = \quad \frac{N\sqrt{Q}}{H_m^{3/4}}$$

Q – Discharge

N - Speed

H_m = Manometric head.

10. What are rotary pump? Give examples.

Rotary pumps resemble like a centrifugal pumps in appearance. But the working method differs. Uniform discharge and positive displacement can be obtained by using these rotary pumps. The various types of rotary pumps one.

   a.   External pump.

   b.   Internal gear pump.

   c.   Vane pump

11. What is meant by cavitation?

It is defined as the phenomenon of formation of vapour bubbles of a flowing liquid in a region where the pressure of the liquid falls below its vapour pressure and the sudden collapsing of these vapour bubbles in a region of higher pressure.

12. What is the effect of cavitation in pumps?

Break down of the machine itself due to severe pitting and erosion of blade surface.

13. Distinguish between pumps in series and pump in parallel.

   1.   If a high head is to developed – the impellers are connected in series.

   2.   If a large quantity of liquid is to developed – the impeller are connected in parallel.

14. What is Air vessel?

An air vessel is a closed chamber containing compressed air in the top portion and liquid at the bottom of the chamber.

15. When will you select a reciprocating pump?

For obtaining high pressure or head and low discharge reciprocating pump is selected.

16. Classify pumps on the basic of transfer of mechanical energy.

   1.   Shape and type of casing.

      a)   Volute casing.

      b)   Vertex casing.

      c)   Casing with guide blade.

2. Working head.

   a) Low head.

   b) Medium head.

   c) High head.

3. Number of stages.

   a) Single stage centrifugal pump.

   b) Multiple stage centrifugal pump.

4. Specific speed.

   a) Low specific speed pump.

   b) Medium specific speed pump.

   c) High specific speed pump.

Unit IV

Pumps

Part A

1. Define slip of reciprocating pump.
2. Mention the working principle of an Air-vessel.
3. Can actual discharge be greater than theoretical discharge in a reciprocating pump?
4. Which factor determines the maximum speed of a reciprocation pump?
5. What are the functions of an air vessel?
6. What is specific speed of a pump? How are pumps classified based on this number?
7. When does negative slip occur?
8. Define slip of a reciprocating pump.
9. When will you select a reciprocating pump?
10. What are Roto dynamic pumps? Give examples.
11. Mention the main components of reciprocating pump.
12. Define "Slip" of reciprocating pump. When does the negative slip occur?
13. When will you select a reciprocating pump?
14. What are rotary pumps? Give examples.
15. Write short notes on types of rotary pumps.
16. Draw a neat sketch. List the components ad briefly explain their functions.
17. Describe the working and principles of a reciprocating pump. List the components and briefly explain their functions.
18. Derive an expression for the work saved in a reciprocating pump by using air vessel.
19. What is indicator diagram?
20. State the main classification of reciprocating pump.
21. Classify pumps on the basis of transfer of mechanical energy.

Part B

1. Show that the work done by a reciprocating pump is equal to the area of the indicator diagram.
2. Classify pumps. Explain the working of a double acting reciprocating pump with a neat diagram.
3. Explain the working principle of reciprocating pump with neat sketch.
4. Define cavitation and discuss its causes, effects and prevention

5. Calculate the work saved by fitting an air vessel for a double acting single cylinder reciprocation pump.

6. The diameter and stroke of a single acting reciprocating pump are 120 mm and 300 mm respectively. The water is lifted by a pump through a total head of 25 m. The diameter and length of delivery pipe are 100 mm and 20 rn respectively. Find out:

    1. Theoretical discharge and theoretical; power required to run the pump if its speed is 60rpm

    2. Percentage slip, if the actual discharge is 2.35 *1/s* and

    3. The acceleration head at the beginning and middle of the delivery stroke.

7. Determine the maximum operating speed in rpm and the maximum capacity in *lps* of a single-acting reciprocating pump with the following details. Plunger diameter = 25 cm, stroke = 50 cm, suction pipe diameter = 15 cm, length = 9 cm, delivery pipe diameter = 10 cm, length = 36 cm, static suction head = 3 m, static delivery head = 20 m, atmospheric pressure - 76 cm of mercury, vapour pressure of w A double acting pump with 35cm bore and 40cm stroke runs at 60 strokes per minute. The suction pipe is 10 m long and delivery pipe is 200m long. The diameter of the delivery pipe is 15cm.The pump is situated at a height of 2.5 m above the sump, the outlet of the delivery pipe is 70 m above the pump. Calculate the diameter of the suction pipe for the condition that separation is avoided. Assume separation to occur at an absolute pressure head is 2.5m of water. Find the Horsepower required to drive the pump neglecting all losses other than friction in the pipes assuming friction factor as 0.02.

8. A single acting reciprocating pump running at 50 rpm, delivers 0.01 m³/s of water. The diameter of the piston is 200 mm and stroke length 400 mm. Determine the theoretical discharge of the pump, coefficient of discharge and slip and the percentage slip of the pump.

9. Explain the working of the working of following pumps with the help of neat sketches and mention two, applications of each.

    1. External gear pump.

    2. Lobe pump.

    3. Vane pump.

    4. Screw pump.

10. Explain the working principle of Gear pump with neat sketch.

Turbines

Part A

1. Differentiate between the turbines and pumps.
2. How are Hydraulic turbines classified?
3. Classify turbines according to flow.
4. What are high head turbines? Give examples.
5. Define hydraulic efficiency of a turbine.
6. The mean velocity of the buckets of the Pelton wheel is 10 m/s. The jet supplies water at 0.7 m³/s at a head of 30 m. The jet is deflected through an angle of 160° by the bucket. Find the hydraulic efficiency. Take Cv = 0.98.
7. Define specific speed.
8. What are the different types of draft tubes?
9. What are the functions of a draft tube?
10. What is a draft tube for Kaplan turbine?
11. Classify turbines according to head.
12. Give an example for a low head turbine, a medium head turbine and a high head turbine.
13. What are reaction turbines? Give examples.
14. Differentiate the impulse and reaction turbine.
15. Draw velocity triangle diagram for Pelton Wheel turbine.
16. Define Hydraulic efficiency.
17. What is draft tube and explain its function?
18. Define the specific speed of a turbine.
19. What is a Draft tube? In which type of turbine it is mostly used?
20. Describe the application of turbine.

Part B

1. Give the comparison between impulse and reaction turbine.
2. Write a note on performance curves of turbine.
3. Write a short note on Governing of Turbines.
4. Derive an expression for the efficiency of a draft tube.
5. With the help of neat diagram explain the construction and working of a pelt on wheel turbine.
6. Show that the overall efficiency of a hydraulic turbine is the product of volumetric, hydraulic and mechanical efficiencies.

7. Obtain an expression for the work done per second by water on the runner of a Pelt on wheel. Hence derive an expression for maximum efficiency of the Pelt on wheel giving the relationship between the jet speed and bucket speed.

8. Explain with the help of a diagram, the essential features of a Kaplan Turbine.

9. A Francis turbine with an overall efficiency of 76% and hydraulic efficiency of 80% is required to produce 150 kW. It is working under a head of 8 m. The peripheral velocity is **0.25 $\sqrt{2gH}$ and** radial velocity of flow at inlet is **0.95 $\sqrt{2gH}$.** The wheel runs at 150 rpm. Assuming radial discharge, determine

    1. Flow velocity at outlet.
    2. The wheel angle at inlet.
    3. Diameter and width of the wheel at inlet.

10. In an inward radial flow turbine, water enters at an angle of 22° to the wheel tangent to the outer rim and leaves at 3 m/s. The flow velocity is constant through the runner. The inner and outer diameters are 300 mm and 600 mm respectively. The speed of the runner is 300 mm. The discharge through the runner is radial. Find the

    1. Inlet and outlet blade angles.
    2. Taking inlet width as 150 mm and neglecting the thickness of the blades, find the power developed by the turbine.

Reg. No.

B.E./B.Tech. DEGREE EXAMINATION, NOVEMBER/DECEMBER 2016.

Third Semester

Mechanical Engineering

CE 6451 – FLUID MECHANICS AND MACHINERY

(Common to Aeronautical Engineering, Automobile Engineering, Mechatronics Engineering, Mechanical and Automation Engineering and Production Engineering, Also common to Fourth Semester Industrial Engineering, Industrial Engineering, Industrial Engineering and Management and Manufacturing Engineering)

(Regulation 2013)

Time: Three hours — Maximum: 100 marks

Any missing data can be suitably assumed

Answer ALL questions.

PART A– (10 x 2 = 20 marks)

1. Write down the effect of temperature on viscosity and gas.

   (a) Calculate the capillary rise in a glass tube of 2.5 mm diameter when immersed vertically in water and (b) mercury. Take surface tension $\sigma = 0.0725 \ N/m$ for water and $\sigma = 0.52 \ N/m$ for mercury in contact with air. The specific gravity for mercury is given as 13.6 and angle of contact = 130°.

2. Find the displacement thickness for the velocity distribution in the boundary layer given by $u/U = 2 \ (y/\delta) - \left(\frac{y}{\delta}\right)^2$

3. Draw the velocity distribution and the shear stress distribution for the flow through circular pipes.

4. State Buckingham's theorem. Why this method is considered superior to Rayleigh's method?

5. Derive the scale ratio for velocity and pressure intensity using Froude model law.

6. What is meant by priming of a centrifugal pump? Why is it necessary?

7. What is the function of air vessel in reciprocating pumps?

8. Explain the type of flow in Francis turbine.

9. What is draft tube?

PART B – (5 x 13 = 65 MARKS)

10. (a) (i) Derive is the Reynold's Transport theorem. (6)

(ii) The dynamic viscosity of an oil use for lubrication between a shaft and sleeve is 6 poise. The shaft is of diameter 0.4 m and rotates at 190 rpm. Calculate the power lost in the bearing for a sleeve length of 90 mm. The thickness of oil film is 1.5 mm. (7)

Or

(b) Derive the Bernouli's equation with the basic assumptions. (13)

11. (a) Derive the Hagen Poiseuille formula for the flow through circular pipes. (13)

Or

(b) Three pipes of 400 mm, 200 mm and 300 mm diameters have lengths of 400 m, 200 m and 300 m respectively. They are connected in series to make a compound pipe. The ends of this compound pipe are connected with two tanks whose difference of water levels is 16 m. If the coefficient of friction for these pipe is same and equal to 0.005, determine the discharge through the compound pipe neglecting first the minor losses and then including them. (13)

12. (a) (i) The pressure difference Δp in a pipe of diameter D and length l due to turbulent flow depends on the velocity v, viscosity μ, density σ and roughness k. Using Buckingham's π theorem, obtain an expression for Δp. (7)

(ii) Define similitude and explain its types. (6)

Or

(i) The pressure drop in an airplane model of size 1/10 of its prototype is 80 N/cm². The model is tested in water. Find the corresponding pressure drop in the prototype. Take density of air = 1.24 kg/m³. The viscosity of water is 0.01 poise while the viscosity of air is 0.00018 poise.

(ii) Derive the five different types of dimensionless numbers. (7)

13. (a) Derive the expression for pressure head due to acceleration in the suction and delivery pipes of the reciprocating pumps.

Or

(b) The internal and external diameter of an impeller of a centrifugal pump which is running at 1200 rpm are 300 mm and 600 mm. The discharge through the pipe is 0.05 m³/s and the velocity of flow is constant and equal to 2.5m/s. The diameters of the suction and delivery pipes are 150 mm and 100 mm respectively and suction and delivery heads are 6 m (abs) and 30 m (abs) of water if the outlet vane angle is 45° and power required to drive the pump is 17 kw. Determine:

i. Vane angle of the impeller at inlet.

ii. Overall efficiency of the pump.

iii. Manometric efficiency of the pump. (13)

14. (a) (i) Describe the efficiencies of a turbine. (6)

(ii) Explain the working of Kaplan turbine. Construct its velocity triangles (7)

Or

(b) The following data is given for Francis turbine: Net Head = 60 m, Speed = 700 rpm, Shaft power = 294.3 kW, η_0 = 84 %, η_h = 93 %, flow ratio = 0.2, breadth ratio = 0.1, outer diameter of the runner = 2 inner diameter of runner. The thickness of vanes occupies 5% of the circumferential area of the runner. Velocity of flow is constant at inlet and outlet and discharge is radial at outlet. Determine :

i. The guide blade angle

ii. Runner vane angle at the inlet and outlet

iii. Diameter of the runner inlet and outlet

iv. Width of the wheel at inlet. (13)

PART C – (1 x 15 = 15)

15. (a) Find the displacement thickness, the momentum thickness and the energy thickness for the velocity distribution in the boundary layer given by $u/U = 2 (y/\delta) - (y/\delta)^2$. (15)

Or

(b) (i) Explain the Reynold's Experiment. (5)

(ii) Derive the Darcy – Weisbach equation for the loss of head due to friction in Pipes. (10)